Deforestation in the Teknaf Peninsula of Bangladesh

Masakazu Tani • Md. Abiar Rahman
Editors

Deforestation in the Teknaf Peninsula of Bangladesh

A Study of Political Ecology

Springer

Editors
Masakazu Tani
Department of Environmental Design,
Faculty of Design
Kyushu University
Fukuoka, Fukuoka, Japan

Md. Abiar Rahman
Department of Agroforestry and
Environment, Faculty of Agriculture
Bangabandhu Sheikh Mujibur Rahman
Agricultural University
Gazipur, Bangladesh

ISBN 978-981-10-5474-7 ISBN 978-981-10-5475-4 (eBook)
DOI 10.1007/978-981-10-5475-4

Library of Congress Control Number: 2017947763

Printed on acid-free paper

This Springer imprint is published by Springer Nature
The registered company is Springer Nature Singapore Pte Ltd.
The registered company address is: 152 Beach Road, #21-01/04 Gateway East, Singapore 189721, Singapore

Foreword

Deforestation and forest degradation are contributing to a slow but steady decline in forest cover on a global scale. Approximately 10–15% of the total carbon dioxide released into the atmosphere every year comes from deforestation and forest degradation. In this volume, Drs. Tani and Rahman and their colleagues provide a comprehensive review of the underlying forces that are driving deforestation in the Teknaf Peninsula in Bangladesh. They describe the deforestation taking place in the Teknaf Peninsula as forest "degradation," a process that takes place over time with no major events but a constant degradation of the forest due to communities that utilize forest land and timber and non-timber forest products to support their livelihoods and eke out a living under challenging economic conditions.

Although globally forest degradation represents 25% of carbon emissions from deforestation and degradation, in Bangladesh, as is the case in many less developed countries, over 80% of emissions come from degradation[1]. The situation in the Teknaf Peninsula in Bangladesh, as is also the case in many developing countries, is that you have local communities encroaching on forested areas often under protection as parks, reserves, and other protected status. With few options to gain a living, these populations often turn to the forest for their sustenance, by clearing forest to plant crops and/or gathering and often selling forest products to meet subsistence needs and generate income. In this volume, the authors provide a description of the problem of forest degradation in the Teknaf Peninsula and go on to provide a well--researched and documented account of the impact and interaction of social, cultural, economic, and environmental factors as drivers of forest degradation.

The value of this book is its comprehensive analysis of the drivers of forest degradation in the Teknaf Peninsula. Forest degradation is the result of the interplay between a number of sociocultural, economic, and environmental factors. Therefore, to understand and address forest degradation, it is important to explore the factors influencing and determining the level of forest degradation. In addition, we need to understand the impact of each factor and the aggregate impact of the suite of factors

[1]Pearson et al. 2017. Greenhouse gas emissions from tropical forest degradation: an underestimated source. Carbon Balance Manager (2017) 12:3. DOI 10.1186/s13021-017-0072-2

driving forest degradation. Understanding the factors and their impact on forest degradation facilitates our ability to come up with solutions to address those factors and reduce degradation.

This book provides detailed information on a large array of social, cultural, economic factors driving forest degradation in the Teknaf Peninsula and their environmental impact that will be helpful in addressing the forest degradation issue there. The methods and analysis also have value on a more general and global scale as an example of the kind of research and analysis that is required to address forest degradation and design feasible solutions to the problem. Three important contributions of the work presented in this volume are as follows:

(1) The volume provides detailed information on the drivers of forest degradation which will help design appropriate solutions to the problem of forest degradation in the Teknaf Peninsula.
(2) The studies provide baseline data on the drivers of forest degradation that can be used to measure sociocultural, economic, and environmental changes in the Teknaf over time in response to any interventions that are implemented and will also provide a framework for monitoring those changes.
(3) The studies and the methodologies employed in this comprehensive work can provide an example and guidance for others in their efforts to better understand and provide solutions to forest degradation in other locations around the globe with similar issues.

The authors should be commended for this important volume that illustrates and provide detailed research on the complicated nature of forest degradation occurring as a result of forests being used to support growing populations on the forest--agricultural border through conversion to agriculture and the harvesting of timber and non-timber forest products to support livelihoods. This comprehensive work provides information that will be key in designing solutions to the problem on the Teknaf Peninsula as well as a framework for researchers and practitioners working under similar circumstances. This will be an important case to watch as the research represented in this volume is translated into practice in our continuing struggle to protect our remaining forest areas while supporting the livelihoods of local communities.

University of Minnesota, USA Dr. Dean Current
April 2017

Acknowledgments

This book is based on a research project conducted since 2009 in the Teknaf Peninsula, Bangladesh. We are grateful to many people who have helped this project in planning, conducting, and supporting the research and publishing this book.

A special word of gratitude is due to Mr. Mohammed Delwar Hossain and Mr. Abu Sayed Mohammed Ismail. They manage local research activities and conduct surveys. Without their contribution, this project would not be possible.

We thank Mr. Momtaz Ahmed Chowdhury, Mr. Hassan Ahmed, and many other local people who helped in collecting data. We appreciate the support of officials of the Forest Department, the Department of Agricultural Extension, and local administrations.

A number of faculty members and students from Kyushu University, Bangladesh Agricultural University, and Bangabandhu Sheikh Mujibur Rahman Agricultural University helped in conducting research and collecting data. We are grateful to all of them.

The research works were in part supported by a grant for environmental research from the Sumitomo Foundation (Grant #093293) from 2009 to 2011, a Grant-in-Aid for Scientific Research (Grant #24401040) by Japan Society for the Promotion of Science (JSPS) from 2012 to 2014, and another JSPS Grant-in-Aid for Scientific Research (Grant #15H02612) from 2015 to 2017. We are grateful for their support.

We are grateful to Professor Dr. Dean Current, Department of Forest Resources, University of Minnesota, for writing the Preface of the book in spite his busy schedule.

Contents

Contributors

K. Asahiro Department of Environmental Design, Faculty of Design, Kyushu University, Fukuoka, Japan

K.K. Islam Department of Agroforestry, Faculty of Agriculture, Bangladesh Agricultural University, Mymensingh, Bangladesh

M. Moriyama Division of Electrical Engineering and Computer Science, Graduate School of Engineering, Nagasaki University, Nagasaki, Japan

A.Z.M. Moslehuddin Department of Soil Science, Faculty of Agriculture, Bangladesh Agricultural University, Mymensingh, Bangladesh

Md. Abiar Rahman Department of Agroforestry and Environment, Faculty of Agriculture, Bangabandhu Sheikh Mujibur Rahman Agricultural University, Gazipur, Bangladesh

M.Z. Rahman Department of Agricultural Extension, Faculty of Agriculture, Bangladesh Agricultural University, Mymensingh, Bangladesh

M. Sakamoto Graduate School of Frontier Sciences, The University of Tokyo, Chiba, Japan

N. Sato Forest Policy Laboratory, Faculty of Agriculture, Kyushu University, Fukuoka, Japan

Masakazu Tani Department of Environmental Design, Faculty of Design, Kyushu University, Fukuoka, Japan

J. Tsuchiya Department of Environmental Design, Faculty of Design, Kyushu University, Fukuoka, Japan

H. Tsuruta Kokusai Kogyo Co. Ltd., Tokyo, Japan

S.M.A. Ullah Department of Agricultural Extension Education, Faculty of Agriculture, Bangladesh Agricultural University, Mymensingh, Bangladesh

About the Editors

Masakazu Tani is a professor of environmental design in the Faculty of Design, Kyushu University, Japan. He also serves as a vice dean of the faculty. His research interests lie primarily in environmental anthropology and environmental design. His main research focus is on the social analysis of environmental problems in South and Southeast Asia. Professor Tani has published articles, reports, and books on the social impacts of arsenic contamination in these regions, both in Japanese and in English. He was awarded the tenth Okita Award for International Development Studies in 2005.

Md. Abiar Rahman is a professor of agroforestry and environment in the Faculty of Agriculture, Bangabandhu Sheikh Mujibur Rahman Agricultural University (BSMRAU), Bangladesh. He served as the head of the Department of Agroforestry and Environment at BSMRAU. His research interests lie in the area of sustainable land use in relation to agroforestry and environment. His main research focus is on the homestead production system and the impacts of climate change on agriculture. He has published 31 articles and supervised 14 MS students in the field of agroforestry and environment. He has attended and presented his works at 19 national and international conferences. In 2015, he was the recipient of an international award for the development of "revolutionary solutions for local food systems," granted by the Italian Agency for Responsible and Ethic Countryside and Agriculture (AiCARE).

Abbreviations

AIGA	Alternative income-generating activities
AIS	Agriculture Information Service
BBS	Bangladesh Bureau of Statistics
BCAS	Bangladesh Center for Advance Studies
BDT	Bangladeshi taka
BFD	Bangladesh Forest Department
BMD	Bangladesh Meteorological Department
CCI	Cambridge Conservation Initiative
CEGIS	Center for Environmental and Geographic Information Services
CWBMP	Coastal and Wetland Biodiversity Management Project
DBH	Diameter at breast height
DMRT	Duncan's multiple range test
ECA	Ecologically Critical Area
FAO	Food and Agriculture Organization of the United Nations
FD	Forest Department (FD)
FGD	Focus group discussion
GoB	Government of Bangladesh
ICEAB	International Conference on Environmental Aspects of Bangladesh
IPAC	Integrated Protected Area Co-management
IUCN	International Union for Conservation of Nature
MSS	Multispectral Scanner
NDVI	Normalized Differential Vegetation Index
NGO	Nongovernmental organizations
NIR	Near-infrared light
NSP	Nishorgo Support Project
RP	Relative prevalence
SF	Social forestry
SPI	Standardized Precipitation Index
SRDI	Soil Resource Development Institute

TM	Thematic Mapper
TWS	Teknaf Wildlife Sanctuary
UAO	Upazila Agriculture Office
UNDP	United Nations Development Programme

Part I
Introduction

Chapter 1
Introduction

Masakazu Tani and Md. Abiar Rahman

Abstract This volume presents studies on deforestation in the Teknaf Peninsula in Bangladesh. The studies analyze the mechanisms and causes of deforestation and explore possible methods of reforestation. The peninsula once had rich forests that harbored a variety of wildlife, including large mammals such as wild boars, deer, and Asian elephants. However, deforestation, which has advanced across the peninsula, has become a critical issue. Currently, clusters of trees in this area are only found in patches of tree plantations and "social" forest plots. Natural forests harbor a very small number of trees, and most areas defined as forests are actually covered only with shrubs and bushes. Although young trees are evidence of regrowth, forests do not appear to be regenerating.

Keywords Betel leaf cultivation • Socioeconomic status • Household survey • Teknaf Peninsula • Major income source

1.1 Tropical Forests and Deforestation

During the last few decades, tropical forests located within 30 degrees of the Equator that were mostly intact prior to the twentieth century have disappeared at a rapid pace. It has been estimated that 20% of all tropical forests, totaling 450 million ha, disappeared between 1960 and 1990 (FAO 2016). Over the last 25 years, commencing from 1990, though the pace of destruction has decreased, deforestation nevertheless remains ongoing in the tropics. Of the three continents where extensive tropical rainforests are distributed, namely South America, Africa, and Asia, the

M. Tani (✉)
Department of Environmental Design, Faculty of Design, Kyushu University, Fukuoka, Japan
e-mail: tani@design.kyushu-u.ac.jp

M.A. Rahman
Department of Agroforestry and Environment, Faculty of Agriculture, Bangabandhu Sheikh Mujibur Rahman Agricultural University, Gazipur, Bangladesh
e-mail: abiar@bsmrau.edu.bd

M. Tani, M.A. Rahman (eds.), *Deforestation in the Teknaf Peninsula of Bangladesh*, DOI 10.1007/978-981-10-5475-4_1

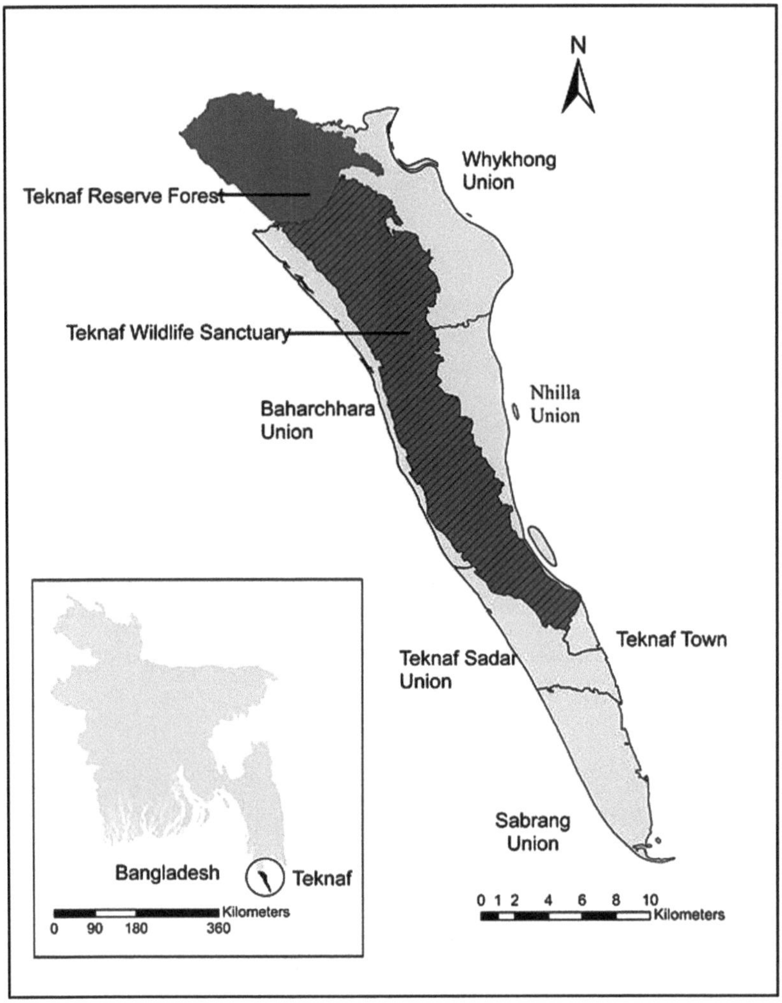

Fig. 1.1 Location of the Teknaf Peninsula and the map of Teknaf Upazila (sub-district) indicating the location of Teknaf Wildlife Sanctuary, a reserve forest, and six unions in the upazila

area of rainforests is the least in Asia, and in recent years (2000–2005), the rate of deforestation has been the highest on this continent (Hansen et al. 2008; Boucher et al. 2011, Chap. 3). As a result, only one-third of Asia's original forests remain intact.

This volume presents studies on deforestation in the Teknaf Peninsula in Bangladesh (Fig. 1.1). The studies analyze the mechanisms and causes of deforestation and explore possible methods of reforestation. The peninsula once had rich forests that harbored a variety of wildlife, including large mammals such as wild boars, deer, and Asian elephants. However, deforestation, which has advanced across the peninsula, has become a critical issue. Currently, clusters of trees in this area are only found in patches of tree plantations and "social" forest plots. Natural forests harbor a very small number of trees, and most areas defined as forests are actually covered only with shrubs and bushes. Although young trees are evidence of regrowth, forests do not appear to be regenerating.

The Teknaf Peninsula is a small and narrow peninsula extending 5–10 km in width and 60 km in length. It is located in the farthest southeastern corner of the country bordering Myanmar, with the two countries separated by the Naf River. The peninsula covers an area of about 400 km^2, which mostly comprises Teknaf upazila (sub-district) with an area of 388.68 km^2. Unqualified references to "Teknaf" in this volume generally refer to the peninsula itself.

Unlike most areas in Bangladesh, the peninsula is predominantly composed of low hills, lacking extensive plain areas. This study on deforestation focuses on the Teknaf Wildlife Sanctuary (hereafter, TWS), which is a reserve forest located in the central part of the peninsula. Occupying a stretch of land covering 116 km^2 TWS accounts for approximately 30% of the area of the sub-district (Fig. 1.1). Another forested area of approximately 40 km^2, which is adjacent to the TWS, is located in the northernmost part of the peninsula (see Chap. 2). Though undeclared, this forest is nevertheless recognized as a protected forest by local offices of the Bangladesh Forest Department. However, deforestation within this latter forest area was not systematically investigated in this study. In its strict sense, deforestation refers to an area that was once forested but has become something other than a forest, for example, a pasture, as a result of loss of trees as well as changes in land use. Forests may also become degraded. Even if there is no change in land use, a forest may become impoverished as evidenced by a decrease in large trees and an increase in shrubs and bushes. This phenomenon is known as forest degradation. Forest conditions may vary along a spectrum between deforestation (in its strict sense) and forest degradation. Changes in land use in the Teknaf Peninsula have been limited. Areas that were once rich forests are mostly still considered forest land and are designated as reserve forests. Therefore, in its strict sense, "deforestation" has not occurred in Teknaf. Nevertheless, a key problem in Teknaf is that the reserve forest in the TWS no longer harbors abundant trees, and new trees do not seem to be maturing despite the supposedly strict protection regime implemented by the Bangladesh Forest Department.

1.2 Drivers of Deforestation

A substantial number of "drivers" of deforestation have been reported from all over the world. These drivers appear to be a multi-layered and inter-linked within a nested structure, collectively contributing to deforestation. Geist and Lambin (2001) proposed a simplified structure of deforestation drivers categorized into two causal types: proximate and underlying.

Proximate causes of deforestation are human activities that directly affect the environment. Underlying driving forces comprise a complex of social, political, economic, technological, and cultural variables that collectively constitute the initial conditions in human-environmental relations (Geist and Lambin 2001). In other words, proximate drivers are direct human actions that alter forests, whereas underlying forces are social/systemic conditions that influence these human actions. Major proximate drivers of deforestation include the expansion of farming fields, cattle pasture, and human settlements; timber logging; and production of pulp and wood fuel, including charcoal extraction. These deforestation drivers are not uniform across the world, with different drivers having differential impacts in specific continents (Boucher et al. 2011). Deforestation in Latin America, which is greatest in terms of its areal extent, is mainly caused by cattle grazing and soybean production. Typically, deforestation entails the conversion of a large area of Amazonian rainforests into agricultural fields and pastureland. Subsequently, roads into the rainforest are constructed, planned, and executed by the state in the name of economic development.

Farming is a major driver of deforestation in Africa, where a substantial area of forests has been cleared to open up agricultural fields for shifting cultivation and for commercial agricultural production. In Asia, and especially in Southeast Asia, oil palm plantations occupy by far the largest proportion of the area of converted forests. Whereas logging has not caused large-scale deforestation in other continents, in Asia, and especially in Southeast Asia, the presence of large quantities of a few commercially important species has led to commercial logging becoming a major driver of deforestation. Moreover, settlements of small subsistence farmers living in and around the forests and fuelwood collection (Boucher et al. 2011) are still important factors in Asia's deforestation (Rudel et al. 2009).

A historical view of deforestation during its early phase from the 1960s to the 1980s reveals that small-scale and state-supported farming in Latin America and Southeast Asia was a key agent of deforestation during this period (Rudel et al. 2009). State governments promoted colonization of rainforests by small and marginal farmers relocating from densely populated areas by building roads penetrating into forested areas, typically located within the Amazon Basin and in the outer islands of Indonesia. From the 1990s onward, land use changes resulting from state-driven colonization of forest areas by small farmers diminished. Changes in land use were caused instead by private enterprises that developed large oil palm plantations in Indonesia and soybean fields and cattle pastures in the Amazon (Rudel et al. 2009).

1.3 Deforestation in Teknaf

During the period from 1970 up to the present, the Amazon lost some 600,000 km^2 of rainforests. Oil palm plantations now occupy 60,000 km^2 of the once-forested area of Borneo. Given that reserve forest in the TWS covers an area of mere 116 km^2, the magnitude of ongoing deforestation in Teknaf is not comparable to that which has occurred in the Amazon and in Southeast Asia. Drivers and underlying processes entailed in deforestation in the TWS also differ from those entailed in large-scale deforestation in the abovementioned regions. Although unauthorized (illegal) logging is said to have been more active in Teknaf in the 1980s, even at their peak, forests in the area were not sufficiently extensive to support a logging industry. Therefore, differing from other major rainforests, enterprise-led and capital-driven deforestation is not—and has not been—evident in this area. Further, the area is not large enough to support extensive farms and pasturage. In fact, plains in Teknaf are not sufficient to support subsistence farming (see Chap. 4). Therefore, the abovementioned major drivers of deforestation such as development of cattle pasturage and soybean fields in the Amazon, and logging for timber and pulp production as well as oil palm plantations in Southeast Asia, have not played a significant role in Teknaf's deforestation.

Because the TWS has been declared and maintained as a reserved forest, these has been no change in land use with minor exceptions entailing illegal encroachers who have individually established homesteads in the forest and patches of agricultural fields of betel plants and sungrass along hill slopes (see Chap. 8). As most of the TWS remains categorized as forest, this is not a case of deforestation in its strict sense. However, the problem is that there are only scattered mature trees in the TWS and trees are not regenerating, even in the absence of the above-mentioned major deforestation drivers. No major industries and enterprises are engaged in extracting trees from the forests. No policies to promote forest colonization have been implemented either by the central or local governments. Consequently, ordinary people living around the forest appear to be only consumers of trees, thereby preventing forest regeneration.

Rudel and Roper (1997) have proposed two models of deforestation. The first, a frontier model, depicts deforestation processes in which coalitions of entrepreneurs, companies, and small farmers are formed "to build roads, establish towns, and develop a region." These actors are organized and deploy private capital as well as assistance procured from the state to open up "the frontier." The other model is an "immiserization" model that portrays peasants and shifting cultivators as the main agents of deforestation. Because of population pressure, and in the absence of economic alternatives, poor farmers are compelled to clear steep hillsides in formerly forested mountainous areas to eke out their subsistence.

The second model appears to be a better fit with Teknaf's deforestation. Although the practice of shifting cultivation is very limited in this area, villagers engage in diverse activities that support their livelihoods in the TWS, despite the fact that the TWS is legally defined as a reserve forest in which all economic activities are generally prohibited. These livelihood supporting activities include establishing new homesteads in the forest, collecting fuelwood and other forest resources, and clearing fields for cultivating various crops and sungrass. Among these activities, fuelwood harvesting and betel leaf cultivation seem to have had major impacts on the state of forests as discussed in subsequent chapters.

Despite the evidently negative impacts of these activities on forests, it is difficult for villagers to curb them. As posited by the immiserization model, because of the absence of economic alternatives, local villagers struggle to subsist on resources obtained from reserve forests to support their livelihoods. A working model for understanding the causes of deforestation in Teknaf may be summarized as follows. The main proximate drivers of deforestation are the daily activities of villagers in the locality of the reserve forests, and the main force underlying these activities is the lack of economic alternatives because of poverty. Because it addresses issues of environmental changes, conflicts, and problems using an approach that links together the social and physical sciences, political ecology provides an appropriate theoretical framework for examining this research question (Painter and Durham 1995; Paulson and Gezon 2005; Robbins 2004). Political ecology analyzes social relations of production and questions of access to and control over resources to better understand the forms of environmental disturbance and degradation and to develop prospects and models for environmental rehabilitation and conservation (Paulson et al. 2005).

Along with this theoretical foundation provided by political ecology, the studies build on the premise that all environmental problems are socially constituted, and that the causes and mechanisms of an environmental problem can only be understood in relation to social factors. Based on the recognition that deforestation in Teknaf is a social problem, the main research objective was to elucidate the relationship between deforestation and poverty. In other words, an underlying assumption was that the poor have little choice but to exploit protected communal resources and deplete forests to secure their survival.

The research activities that constituted the overall study were organized around two themes. The first entailed developing an understanding of the mechanisms of deforestation, and the second focused on advancing a potential methodology of reforestation. Reflecting these themes, this volume is organized into four parts and 15 chapters.

Part I presents the background of the overall study. Chapter 1, this chapter, introduces the problem of deforestation in Teknaf. Chapter 2 describes the physiography, climate, forests, and people in the Teknaf Peninsula. These descriptions intend to provide descriptive and detailed information of the study area and to set the framework of studies in the following chapters in Part II and Part III.

The chapters in Part II present analyses of the factors currently contributing to deforestation processes in Teknaf. Chapters 3, 4, 5 and 6 engage with different aspects of betel leaf cultivation. Unlike deltaic areas characterized by fertile top soils and abundant water supplies that comprise most of Bangladesh, the southern half of the Teknaf Peninsula is characterized by hilly land with limited plain areas and poor-quality soils. Therefore, areas suitable for wet rice farming are limited at best. Because of these conditions, the betel plant, which can be grown on sloping ground, has become an important cash crop that compensates for the paucity of rice cultivation in this area. However, to produce high-quality betel leaves for sale, facilities for providing the plants with shade are required during their growth. These facilities are mainly constructed using forest resources, as discussed in subsequent chapters.

Chapter 3 discusses the impacts of betel leaf cultivation on forests, demonstrating that this practice is a major cause of deforestation in this area. Chapter 4 examines the socioeconomic status of betel leaf farmers, finding that although they are not situated on the lowest rungs of the socioeconomic ladder, they tend to belong to poorer farmer segments. Chapter 5 analyzes the economic impacts of betel leaf cultivation on farmers. Chapter 6 combines analyses of satellite images and of the findings of a socioeconomic survey to reveal the relationship between the progression of betel leaf farming and deforestation.

Because almost all of the cooking fuel in this area consists of fuelwood, substantial quantities of fuelwood are procured from nearby forests. Therefore, Chap. 7 presents an analysis of the collection, distribution, and consumption of fuelwood. This study includes analyses of fuelwood supply chains, controlled measurements of fuelwood uses within households, and intermediate fuelwood vendors as well as market-based vendors.

Along with the use of forest products for betel leaf cultivation and fuel, human settlements in the forest also have more direct impacts on the state of forests. Although most forests in the study area are designated as reserve forests in which all human activities are banned, in principle, an exception to this rule applies to people who have special rights to live in forests. Based on detailed household surveys conducted in settlements located within the boundaries of a reserve forest, Chap. 8 presents an analysis of communities in this area. It attempts to distinguish between encroachers and those who have legal rights to reside within the forest and compares their respective attributes.

Chapter 9, which is the last chapter in Part II, focuses on the Rohingyas. The Rohingyas are a Muslim minority group based in Myanmar who became refugees after fleeing the country to escape from oppression by the non-Muslim majority. Although the United Nations High Commissioner for Refugees has established two refugee camps in the Teknaf Peninsula, many more refugees live within local communities. Because they do not have any property, resources, and jobs, they must somehow secure their survival, often by exploiting "free" natural resources. Therefore, they are often blamed as being a major cause of deforestation in this area. Chapter 9 discusses the livelihoods of the Rohingyas in Teknaf, especially in connection with their forest-related activities, based on a study of a large Rohingya settlement.

Part III of this volume comprises chapters that deal with issues potentially related to reforestation. The social forestry programs implemented by the Bangladesh Forest Department have been the most organized effort conducted to date to reforest this area. A social forestry program entails designating an area as a plot and selecting a group of local beneficiaries who are responsible for maintaining the forest and who receive a portion of profits derived from the forest. Chapter 10 describes the current conditions of all 37 social forestry plots in the study area. Their conditions were evaluated based on physical observations as well as tree censuses conducted within each plot. Chapter 11 analyzes the beneficiaries and individuals involved in developing and maintaining social forestry plots in terms of their roles and power.

The last three chapters in Part III discuss homestead gardens and forests. Homesteads are seen to be a source of tree production as well as of other less durable resources. Chapter 12 presents an analysis of homesteads, viewed as production units, based on the findings of a questionnaire administered among the respective households and reports on the distribution of plants within homesteads. This chapter also presents a detailed analysis of trees within homesteads based on complete tree censuses that recorded all of the trees found in individual homesteads. Chapter 13 identifies the ecosystem benefits provided by homesteads through their supply of forest products, ensuring production, conserving biodiversity, and generating incomes. Chapter 14 examines the variability in homesteads maintained by four different ethnic groups, Bengali, Chakma, Rakhaine, and Rohingya, in this area. The analysis finds that different groups have different emphasis on plants grown in homestead gardens, but, at the same time, that all homesteads represent the wise use of small areas of land to reduce pressure on forests.

Part IV, Chap. 15 presents the study's conclusions. Specifically, it discusses the relationship between deforestation and poverty and presents an assessment of the drivers affecting Teknaf's deforestation. It also evaluates current reforestation efforts and their prospects. Last, an attempt is made to form a prospectus of further studies based on the results of this study.

References

Boucher D, Elias P, Lininger K, May-Tobin C, Roquemore S, Saxon E (2011) The root of the problem: What's driving tropical deforestation today? Union of Concerned Scientists. Online at http://www.ucsusa.org/sites/default/files/legacy/assets/documents/global_warming/UCS_RootoftheProblem_DriversofDeforestation_FullReport.pdf

FAO (2016) Global forest resources assessment 2015. Food and Agriculture Organization of the United Nations, Rome. Online at http://www.fao.org/forest-resources-assessment/en/

Geist H, Lambin E (2001) What drives tropical deforestation? LUCC report series no. 4. University of Louvain, Belgium

Hansen M et al (2008) Humid tropical clearing from 2000 to 2005 quakified by using multitemporal and multiresolution remotely sensed data. Proc Natl Acad Sci U S A 105:9439–9444

Painter M, Durham W (eds) (1995) The social causes of environmental destruction in Latin America. University of Michigan Press, Ann Arbor

Paulson S, Gezon L (2005) Political ecology across spaces, scales and social groups. Rutgers University Press, New Brunswick

Paulson S, Gezon L, Watts M (2005) Politics, ecologies, genealogies. In: Paulson S, Gezon LL (eds) Political ecology across spaces, scales and social groups. Rutgers University Press, New Brunswick, pp 17–40

Robbins P (2004) Political ecology: a critical introduction. Blackwell, Malden

Rudel T, Roper J (1997) The path to rain forest destruction: crossnational patterns of tropical deforestation, 1975–90. World Dev 25(1):53–65

Rudel T, Defries R, Asner G, Laurance W (2009) Changing drivers of deforestation and new opportunities for conservation. Conserv Biol 23(6):1396–1405

Chapter 2
Physiography, Forests, and People in Teknaf

A.Z.M. Moslehuddin, Md. Abiar Rahman, S.M.A. Ullah, M. Moriyama, and Masakazu Tani

Abstract The Teknaf peninsula, which is located in the farthest southeastern corner of Bangladesh, has a diversified physiography exhibiting hills, piedmont plains, tidal floodplains, and beaches. The climate is subtropical, with temperatures ranging seasonally from 15 °C during the winter to 33 °C during the summer and very heavy rainfall (about 4000 mm). The peninsula, which experiences frequent cyclones, is vulnerable to climate change. More than 30% of the area consists of forests that are highly degraded as a result of various human activities as well as climate change such as frequent cyclone. Poverty is one of the major issues in the peninsula, with approximately 38% of the population living below the poverty line. Livelihoods center on agriculture, forest-related activities, and collection of marine resources. Although human settlements have existed in this area from ancient times, after 1990, extensive settlement has occurred in and around the forests. Apart from Bengalis, other ethnic groups living in the peninsula include the Chakma, Rakhine, and Rohingya groups.

Keywords Land use • Land types • Climate • Reserved forest • Livelihood • People

---

A.Z.M. Moslehuddin (✉)
Department of Soil Science, Faculty of Agriculture, Bangladesh Agricultural University, Mymensingh, Bangladesh
e-mail: abunazia@yahoo.com

M.A. Rahman
Department of Agroforestry and Environment, Faculty of Agriculture, Bangabandhu Sheikh Mujibur Rahman Agricultural University, Gazipur, Bangladesh
e-mail: abiar@bsmrau.edu.bd

S.M.A. Ullah
Department of Agricultural Extension Education, Faculty of Agriculture, Bangladesh Agricultural University, Mymensingh 2202, Bangladesh
e-mail: asiksm@gmail.com

M. Moriyama
Division of Electrical Engineering and Computer Science, Graduate School of Engineering, Nagasaki University, Nagasaki, Japan
e-mail: matsu@rsirc.cis.nagasaki-u.ac.jp

M. Tani
Department of Environmental Design, Faculty of Design, Kyushu University, Fukuoka, Japan
e-mail: tani@design.kyushu-u.ac.jp

M. Tani, M.A. Rahman (eds.), *Deforestation in the Teknaf Peninsula of Bangladesh*, DOI 10.1007/978-981-10-5475-4_2

2.1 Introduction

Teknaf upazila (an upazila is an administrative unit of local government in Bangladesh) is located in Cox's Bazar District in the southeastern part of Bangladesh. It is bordered by Ukhia upazila to the north, the Naf River and Myanmar to the east, and the Bay of Bengal along its western and southern borders. Because three sides of this area, which includes Teknaf's administrative and forest areas, are surrounded by water bodies, it is referred to as the Teknaf peninsula. However, in this book, we have adopted three distinct terms as follows: Teknaf peninsula (the entire area), Teknaf upazila (the administrative area), and Teknaf Wildlife Sanctuary (the forest area). The center of Teknaf upazila is located at a distance of about 80 km from the district headquarters. It comprises six unions, 12 mouzas (several villages typically form a single mouza), six unions, and one pourashava (pourashava is the town or center of the upazila and termed as Teknaf town also) comprising 147 villages. It has a population of 265,717 distributed over an area of 388.68 km^2. The literacy rate in this area is 19.72%, which is well below the national average (51.8%). The area exhibits diverse land types, with wet, forested, hilly, coastal, and flat land found together within a narrow stretch of land. The two main agricultural crops cultivated in the area are the *Aman* rice variety and betel leaf. A locality's climate is one of the factors determining the type and density of vegetation. This area has a subtropical climate characterized by a relatively high amount of annual rainfall (more than 4000 mm) and an average temperature of 25.5 °C. The physiographic and climatic conditions of the area are conducive to the growth of forests. At one time, forests were dominant. However, although 41% of the area comprises forests, these are highly degraded as a result of extensive anthropogenic activities, as well as natural phenomena. The area is also characterized by cultural variability with settlements of Bengalis, Chakmas, and Rohingyas living in Teknaf upazila. Collection of forest resources, fishing, farming, business, and labor are the main livelihood activities in the area. This chapter describes the physiography, climate, forests, and people of the Teknaf peninsula under separate subsections.

Table 2.1 provides general statistics for the area. More than 0.26 million people live in an area covering 388.68 km^2 in Teknaf upazila. Teknaf's population density is 549 persons per km^2, which at almost half the national average (BBS 2015) is low compared with that of other regions in the country. Whereas about 50% of households' landholdings are apportioned for agriculture in Bangladesh as a whole, in Teknaf, agriculture accounts for about 35% of households' landholdings. Among farm families in Teknaf, those with small farms make up the highest proportion of farmers (31.77%), followed by marginal farmers (28.40%), those with medium-sized farms (20.41%), the landless (16.30%), and those with large farms (3.12%). A total of 14,602 ha of land are under forest of which 11,615 ha have been declared as Teknaf Wildlife Sanctuary (TWS).

In Teknaf, forests play a vital role in sustaining the subsistence livelihoods of local people. Most parts of Teknaf consist of highland areas, with only a few areas being lowlands, indicating that the area is not flood prone. Unlike in other coastal areas in Bangladesh, salinity does not pose a major problem in Teknaf, with a little over 8%

Table 2.1 General statistics of Teknaf

Parameter	Value
Population	265,717
Population density per km^2	549
Total area (km^2)	388.68
Literacy rate (%)	26.7
Total farm family	18,430
Large (%)	3.12
Medium (%)	20.41
Small (%)	31.77
Marginal (%)	28.40
Landless (%)	16.30
Forest land (ha)	14,602
Highland (ha)	3842
Medium highland (ha)	5800
Medium lowland (ha)	2736
Lowland (ha)	750
Saline area (ha)	32,425
Total cultivable land (%)	5.5
Cropping intensity (%)	136

Source: UAO-Teknaf 2016

of land being saline. Only 5.5% of the land is under cultivation, having a cropping intensity of 136%, which is well below the national average of 193% (AIS 2017). Rahman et al. (2014) reported that agriculture in Teknaf showed a significant increase after 1990 following extensive clearing of forests to create a huge settlement. Our own recent study (the data are unpublished) shows that agricultural productivity is lower in Teknaf compared with productivity in many other parts of Bangladesh because of the use of conventional crop varieties and the lack of know-how. As a result, the upazila is acknowledged to be a food-deficit area (UAO-Teknaf 2016).

2.2 Physiography of Teknaf

The term physiography includes the combination of the geological material in which particular kinds of soils have formed and the landscape on which they occur. Bangladesh has been divided into 34 physiographic units and subunits (FAO-UNDP 1998). Physiographically, the Teknaf peninsula mainly comprises: (i) hills, (ii) piedmont plain, (iii) tidal floodplain, and (iv) beach, with a minor area of coral beach. These areas cover approximately 80% of the total landmass (SRDI 2001):

1. The hill area covers 14,602 ha, accounting for 41.8% of the total area. According to its height and morphology, the hill area can be divided into two types: medium-high hills and medium-low hills. The area of medium-high hills is approximately

6940 ha, and the heights of the hills range from 150 to 300 m. The slopes of these hills range from steep (30–50% slopes) to excessively steep (>70% slopes). The hill ranges, which are interrupted by streams and valleys, are oriented from north to south, and they developed over sedimentary rocks. The area of medium-low hills, which developed over soft sedimentary rocks, is 7436 ha, and the hill heights are less than 150 m. The slopes range from excessively steep by steep to slightly steep (<5% slopes). Erosion is very common in steep to very steep areas, while it is less common in flatter areas. Deep soils only developed in flatter areas. Narrow and broad ridges are found alongside the low hills.

2. The piedmont plain covers 3034 ha (8.6% of the total area). It is situated alongside the hills, mainly on their western side, but is found sporadically on the eastern and south sides of the hills. The landscape is a nearly level high ridge, which is subject to flash floods during the rainy season.
3. The tidal floodplain, which runs from north to south through the peninsula, comprises 6838 ha of land (19.57% of the total area). This is located between the hills and the Naf River (on the eastern side of the hills). The area consists of broad, high, and low ridges and depressions. Numerous canals divide the landscape, some of which are subjected to tidal flooding. During the rainy season, most of the areas become mildly inundated with rainwater and occasionally suffer flash floods during heavy rainfall.
4. Beaches cover 9.03% (3155 ha) of the total area, and they lie on the west side of the peninsula along the sea. The landscape is mostly flat with some undulating relief consisting of sandy soil. The coral beach is a minor area (1%) that is located approximately 12 km from the mainland. It is located on St. Martin's Coral Island. The landscape consists of very gently undulating old beach ridges and inter-ridge depressions, which are surrounded by sandy beaches.

2.2.1 Land Use and Topography

Teknaf upazila land use map (Fig. 2.1) shows the different types of occupancy of the area. Forest covers nearly 15,000 ha areas which is 39% of the total area. Among the forest area, 11,610 ha area is declared as Teknaf Wildlife Sanctuary (TWS). Paddy and other agricultural practices such as watermelon, betel leaf, and chili cover a significant portion of land use. Teknaf upazila comprises nearly 57,000 households, and most of the households have homestead vegetation around. Salt fields are seen in the eastern side of the Teknaf upazila where people harvest saline water from the Naf River and dry the water to collect salt. Among the other land use, there are beaches and waterlogged lands.

Figure 2.2 represents the topography of Teknaf peninsula by using contour lines. Contour lines are lines drawn on a map connecting points on equal elevations. This map was prepared by using 100-m contour intervals from 5-m resolution digital elevation model which was made from the satellite of ALOS images that was taken from 2006 to 2011 all images, and it was made by NTT DATA Corporation in

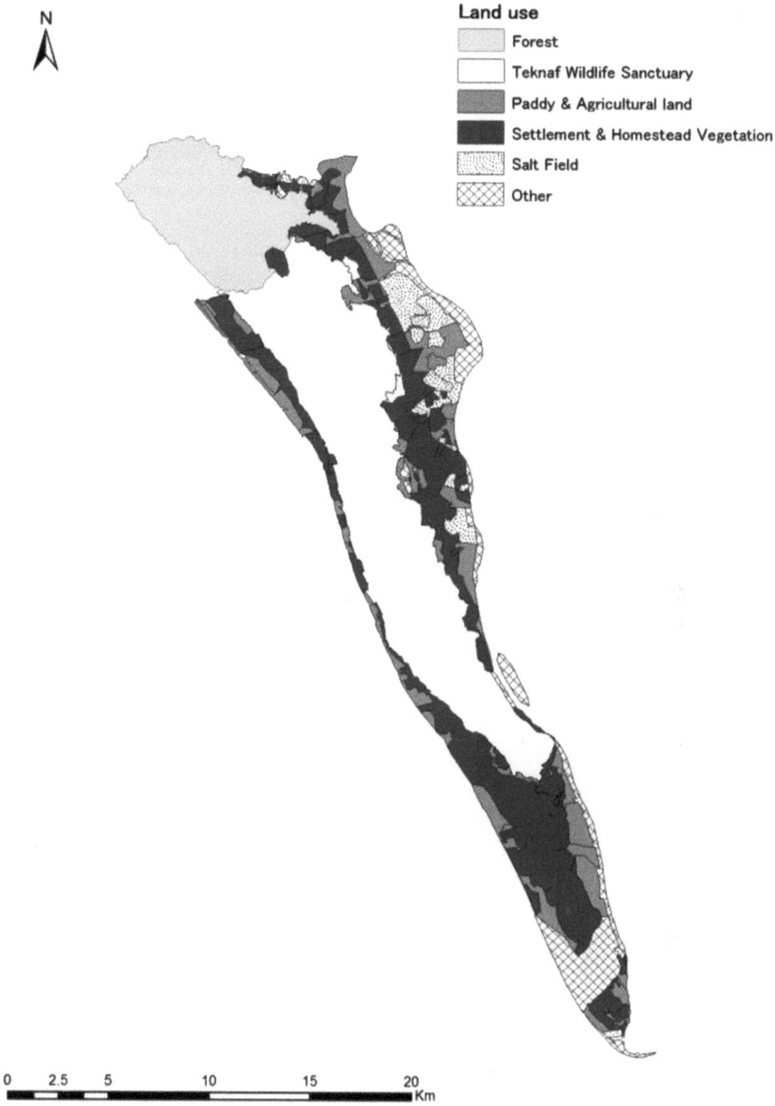

Fig. 2.1 Land use of Teknaf upazila

September 2015. So, this 100-m contour interval map shows contour lines for every 100 m of elevation lines from sea level at 0, 100, and 200 m. This concentric closed contours indicating that it is a hilly area with uneven surface but elevation is below 300 m. Also this contours line space indicating that the elevation is not changing much and east side slope is steeper than the west side.

Figure 2.3 represents the topography of Teknaf peninsula by using shaded relief. Relief is the difference in elevation (or height) between parts of the Earth's surface,

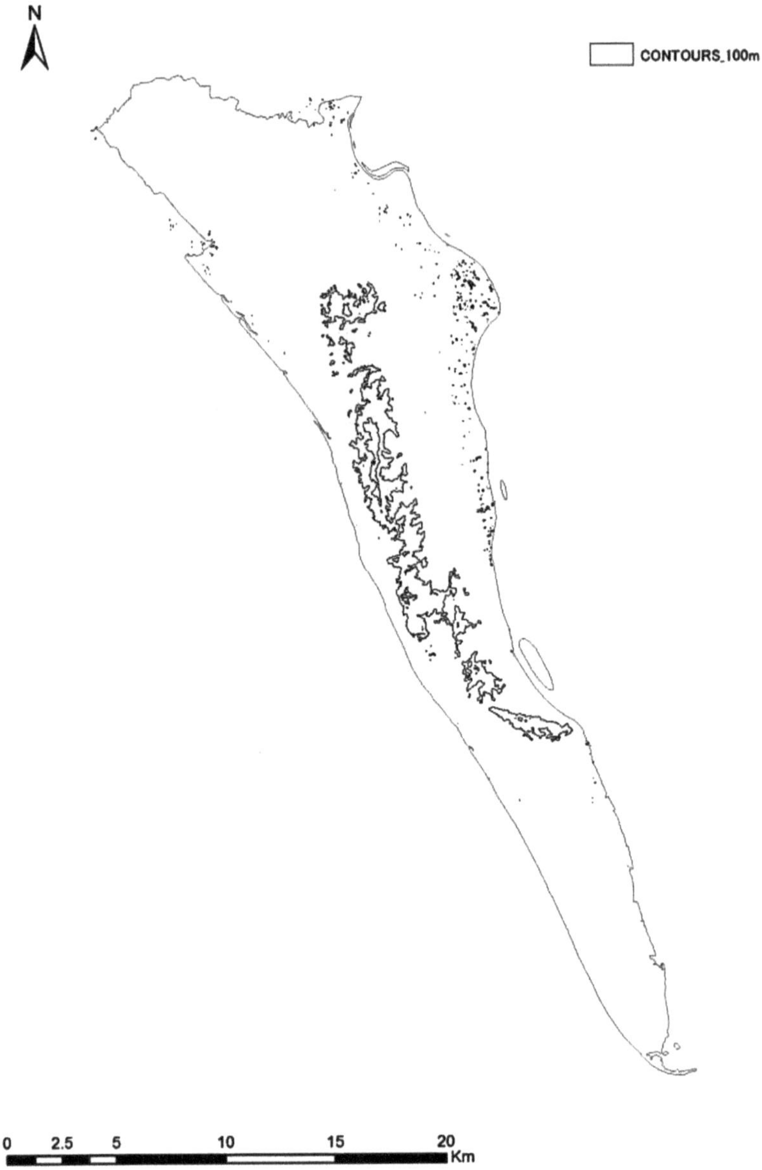

Fig. 2.2 Topographic map (by contour line)

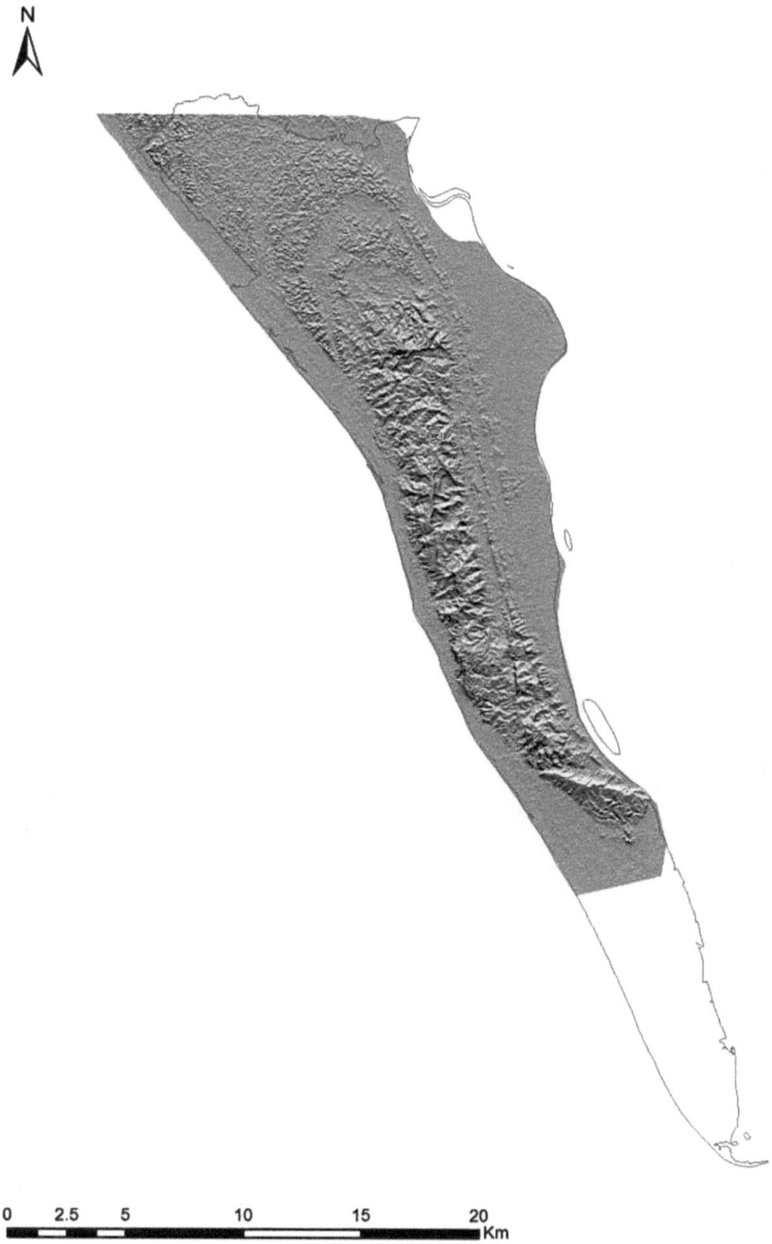

Fig. 2.3 Topographic map (by shaded relief)

and shaded relief maps show features on the surface, such as mountains, valleys, plain land, etc. This is a 5-m resolution digital elevation model (DEM) which was made from the satellite of ALOS images that was taken from 2006 to 2011 all images, and it was made by NTT DATA Corporation in September 2015. From this map, areas of middle Teknaf appear with rough surface that indicates steep slopes and mountains, east and west side areas are smooth on map that are flat plain land, and the upper areas of Teknaf are not as smooth as flat plains or not much as rough as steep slopes, but it is a highland area. The lower part of Teknaf peninsula is flat plain area which is not covered by this map.

2.3 Climate

The Teknaf peninsula is characterized by a subtropical climate, with temperatures ranging from 15 to 33 °C in winter (January) and summer (May), respectively. There is frequent heavy rainfall, and total annual rainfall is approximately 4000 mm; however, rainfall varies widely throughout the year. The climatic conditions of the area will be described subsequently.

2.3.1 Database and Calculations

The normal climatic parameters at a given station at any scale can be assumed to be the mean over a 30-year period. Using this criterion, a dataset from 1984 to 2013 was collected from the Bangladesh Meteorological Department (BMD 2014) and analyzed to characterize the climate of the Teknaf peninsula. Mean monthly maximum and minimum temperatures were derived by averaging daily maximum and minimum temperatures. Bangladesh has a subtropical monsoon climate characterized by wide seasonal variations in rainfall, temperature, and humidity; therefore, to determine seasonal variability, the monthly data were separated into three seasons: summer (March to June), monsoon (June to October), and winter (October to March).

The standardized precipitation index (SPI) was developed to identify and monitor droughts from long-term (≥30-year) monthly precipitation records (McKee et al. 1993). It assesses anomalous and extreme precipitation amounts by giving them a numeric value, thereby enabling the tracking and comparison of meteorological droughts across areas with different climates. The SPI calculation for any location is based on the long-term precipitation records for the desired time period. This long-term record is fitted to a probability distribution, which is then transformed into a normal distribution (Edwards and McKee 1997), and SPI is reflective of the number of standard deviations that an observed value digresses from the long-term mean.

Table 2.2 Classification of drought based on SPI

SPI values	Drought category
2.00 and above	Extremely wet
1.50 to 1.99	Very wet
1.00 to 1.49	Moderately wet
−0.99 to 0.99	Near normal
−1.00 to −1.49	Moderately dry
−1.50 to −1.99	Severely dry
−2.00 and less	Extremely dry

$$SPI = \frac{Xi - \overline{X}}{\sigma}$$

where, Xi, $\overline{X}$, and σ are *i*th year precipitation, long-term mean of precipitation, and the standard deviation of the mean, respectively.

Hayes et al. (1999) suggested drought classification based on the SPI values, in which positive values indicate wet conditions and negative values signify dry periods (Table 2.2). In this study, an SPI value ≤ -1 is considered to be drought initiation as described by Wu et al. 2007.

2.3.2 *Average Temperature and Rainfall*

The long-term rainfall and temperature in the Teknaf peninsula are shown in Table 2.3. The average annual rainfall was 4066.3 mm. July and January were the wettest and driest months, accounting for 26.8% and 0.1% of the annual rainfall, respectively. Among the seasons, the monsoon rainfall was 3380.3 mm (83.1% of the annual rainfall), followed by that in summer (352.2 mm, 8.7%) and winter (333.8 mm, 8.2%).

The mean maximum and minimum temperatures were 30.2 and 22.1 °C, respectively. The hottest and coldest months were May (32.2 °C) and January (14.9 °C), respectively. The seasonal variations of the maximum temperature showed that summer was the warmest time (temperatures ranged from 30.9 to 32.2 °C), followed by the monsoon season (29.9–30.8 °C) and winter (27.4–31.4 °C). Similarly, the minimum temperature was higher during summer (ranged from 20.6 to 25.4 °C), followed by the monsoon season (25.2–25.5 °C) and winter (14.9–21.0 °C).

Table 2.3 Variation of mean monthly rainfall and temperatures on the Teknaf peninsula from 1984 to 2013

Season	Rainfall (mm)	Temperature (°C)	
		Maximum	Minimum
Winter	3.4	27.4	14.9
	13.2	29.0	16.9
Summer	13.7	30.9	20.6
	62.8	32.1	24.1
	275.5	32.2	25.4
Monsoon	952.0	30.4	25.5
	1090.7	29.9	25.3
	907.6	30.0	25.3
	430.1	30.8	25.2
Winter	231.8	31.4	24.3
	76.7	30.3	21.0
	8.9	28.2	16.9
Average	–	30.2	22.1
Total	4066.3	–	–

2.3.3 *Climatic Trends 1984–2013*

Monthly distribution patterns of rainfall and temperature are shown in Fig. 2.4. Long-term data show that more than 4000 mm rainfall occurs annually, which is higher than many other locations of Bangladesh. June, July, and August are the rainiest months, though the maximum rainfall (1090 mm) occurs in the month of July. These 3 months contribute about 73% of the total rainfall in a year. Negligible amount of rainfall occurs during the months from December to March, which is considered as dry period. The maximum temperature varies from 27.3 to 32.2 °C, while minimum temperate lies between 25.4 and 14.9 °C. May and January are the warmest and coldest months, respectively.

The long-term trend in annual rainfall showed a slight increase (2.2 mm per year) over time. However, the trend for seasonal variation was remarkable, as the monsoon rainfall increased by 4.6 mm per year, while rainfall decreased during summer (1.47 mm per year) and winter (0.93 mm per year) (Fig. 2.5). The SPI values showed frequent changes in dry and wet conditions in the study area (Fig. 2.6). Although total rainfall was static, dry and wet conditions prevailed because of an uneven rainfall distribution. However, droughts have occurred more frequently in recent years.

The maximum temperature showed a slight increasing trend, while the minimum temperature was almost static over time (Fig. 2.7). The maximum temperature increased over time, irrespective of season. The per annum rate of increase was highest in winter (0.014 °C), followed by summer (0.003 °C) and the monsoon season (0.0015 °C), whereas the average (all season) rate of increase was 0.006 °C. In contrast, the minimum temperature showed increasing and decreasing trends over time because of seasonal variations. The minimum temperature for the monsoon

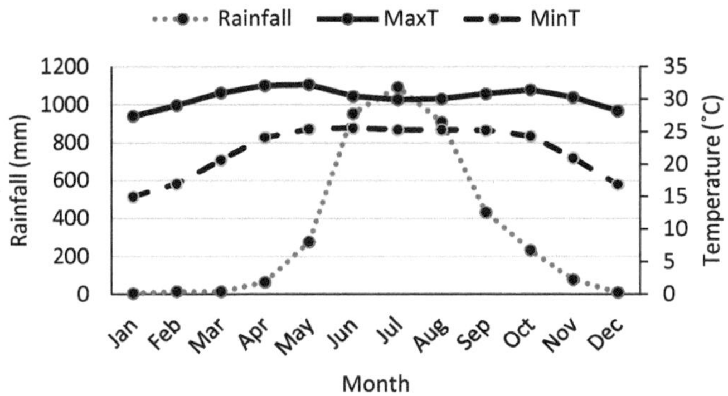

Fig. 2.4 Monthly distribution patterns of rainfall and temperatures based on long-term data (1984–2013) in the Teknaf peninsula

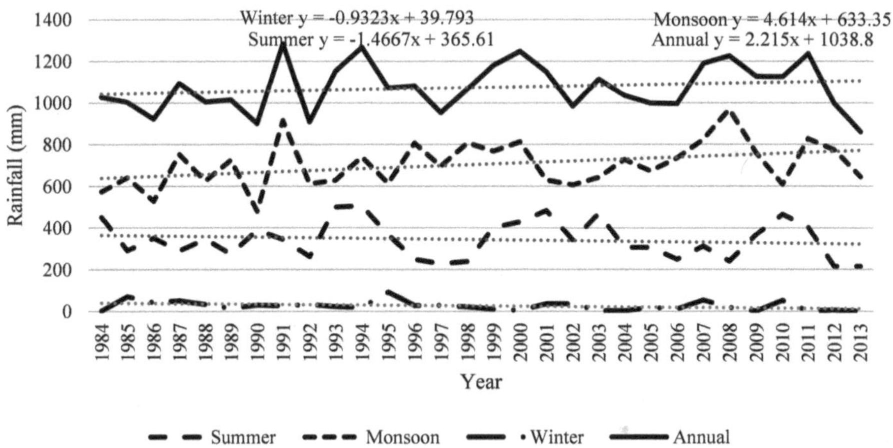

Fig. 2.5 Annual and seasonal rainfall variations on the Teknaf peninsula

season increased by 0.027 °C, while it decreased by 0.013 and 0.015 °C during summer and winter, respectively (Fig. 2.8).

2.3.4 Natural Disasters

As in other coastal areas, Teknaf is prone to frequent cyclones. In general, cyclones are frequent before and after the monsoon season. Based on long-term data, it appears that cyclones occurred most frequently during the months of May, October, and November (Fig. 2.9). Dasgupta et al. (2010) observed that only 11 cyclones hit the coast of Bangladesh from 1901 to 1957, while there were 55 cyclones from 1957 to 2009. Thus, during the last 52 years, the number of cyclones

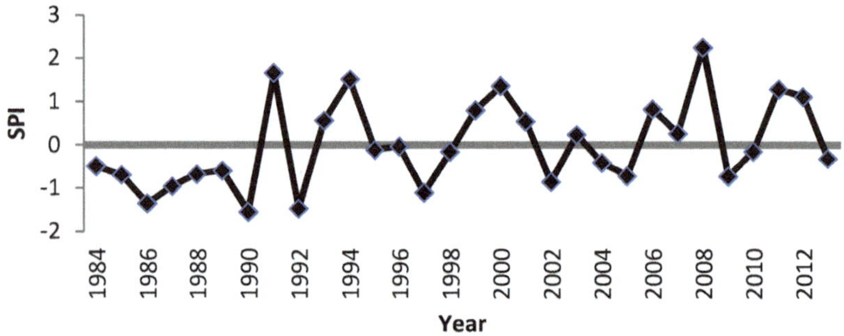

Fig. 2.6 Long-term (1984–2013) annual SPI in Teknaf indicating the frequency of dry and wet conditions

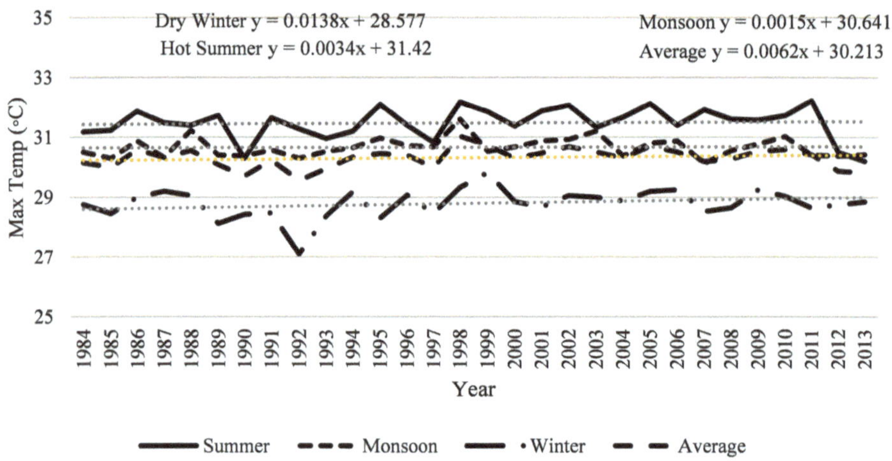

Fig. 2.7 Monthly average and seasonal variations of maximum temperature on the Teknaf peninsula

hitting coastal areas of Bangladesh was fivefold greater than that during the previous 57 years.

In terms of cyclone-related casualties, cyclone intensity is more important than frequency. A long-term dataset showed that cyclone intensity varied (Fig. 2.10), but on average, a severe cyclone strikes Bangladesh every 3 years (GoB 2009a). A study of cyclone frequency found that the Bay of Bengal has the second highest cyclone frequency, experiencing approximately 20 cyclones per decade. According to the Intergovernmental Panel on Climate Change, future tropical cyclones will become stronger, with faster wind speeds (Pender 2010). Because of high ocean temperatures, the frequency and intensity of such storms will increase (Williams 2002).

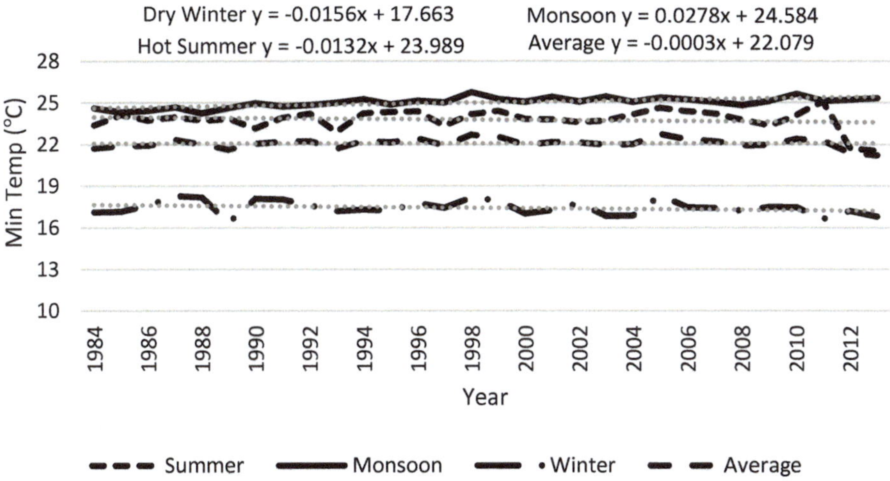

Fig. 2.8 Monthly average and seasonal variations of minimum temperature on the Teknaf peninsula

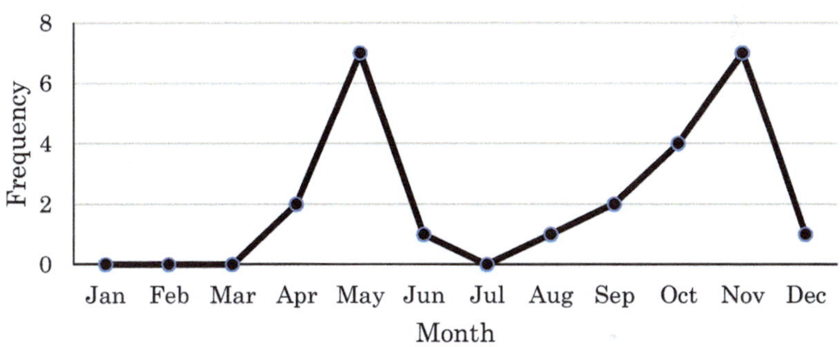

Fig. 2.9 Frequency of storms in the coastal area of Bangladesh (Modified after Alam et al. 2003; GoB 2009a)

Long-term climatic data indicate an increasing trend in monsoon rainfall, which increased by 66%, while rainfall during the dry season decreased by 29% in recent years compared with that of 30 years ago. The maximum temperature showed a slightly increasing trend over time, while the minimum temperature remained unchanged. However, although the annual rainfall showed an increasing trend, rainfall was not well distributed throughout the year. In recent years, there were prolonged droughts during the dry season and high rainfall during the monsoon season. This patchy rainfall is not good for forest resource conservation and agriculture production. Climate change is likely to accelerate forest degradation.

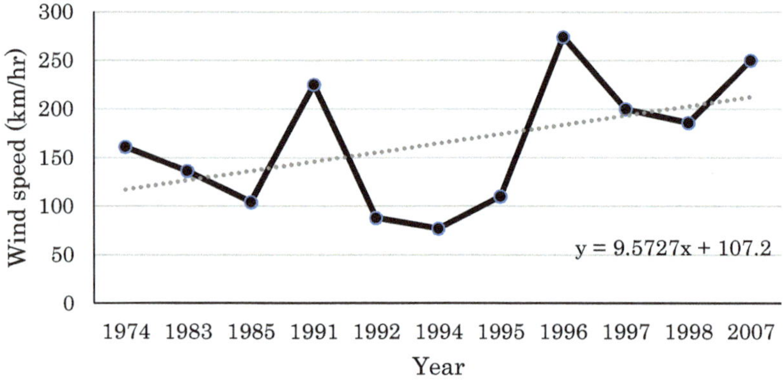

Fig. 2.10 Wind speed of cyclones in the coastal area of Bangladesh (Modified after Dasgupta et al. 2010; GoB 2009a)

2.4 Forest

2.4.1 History and Boundary

Most forested areas in the southern half of the Teknaf peninsula became a reserve forest in 1907 (Belal 2013). The same area was declared the Teknaf Game Reserve (TGR) in 1983 under the Bangladesh Wildlife Act of 1974 (Alam et al. 2012; GoB 2009b) to preserve wild elephants and other key wildlife and to facilitate game hunting for sport hunters (Bari and Dutta 2004). As per the provision of the Act, all of the wildlife species found in the game reserve shall be protected and preserved. Elephant conservation has been given high importance, as they are considered to be endangered within their total ranges in Asia and Bangladesh. In reality, the TGR was not used for game hunting because it did not have enough game animals. Therefore, its status was changed to the Teknaf Wildlife Sanctuary (TWS) in 2010 (BFD 2014).

The TWS is a part of the Teknaf peninsula (Fig. 2.11), and it is located in the southeastern corner of Bangladesh. The geographical position of the TWS is 20°52′–21°09′N and 92°08′–92°18′E (Rosario 1997). The TWS is roughly 28 km from north to south and 3–5 km from east to west. The Naf River and estuary are east of the forest, and they form the border between Bangladesh and Myanmar. The west side is bordered by the Bay of Bengal.

The TWS is one of the largest protected areas in Bangladesh. The forest is managed by the Divisional Forest Office of Cox's Bazar (South) through three range offices (Teknaf, Whykhong, and Shilkhali ranges) and 11 forest beats (Teknaf, Mochoni, Hnila, Madhya Hnila, Rajarchara, Mathabhanga, Shilkhali, Shaplapur, Whykhong, Raikheong, and Monkhali). Approximately 45 Bangladesh Forest Department (BFD) officers and staff members work in the forest reserve (BFD 2006). The total area of the forest is 11,610 ha (Green 1987).

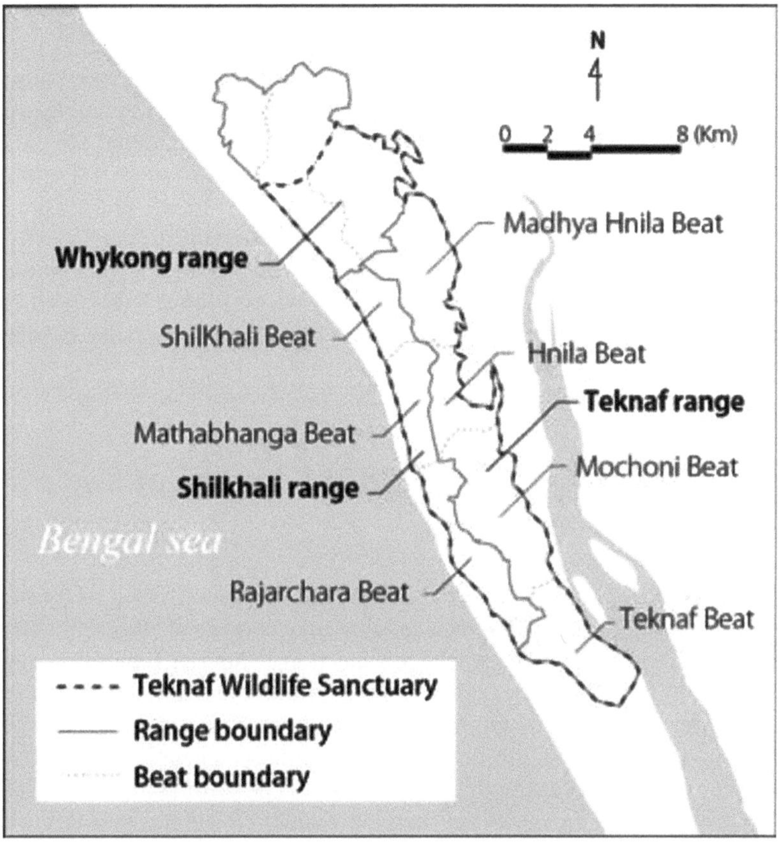

Fig. 2.11 Map of the Teknaf peninsula showing the TWS and range and beat boundaries

2.4.2 Characteristics of the Forest

The TWS is a part of a linear hill range, which consists of gently sloping to rugged hills and cliffs running down the central part of the peninsula. The range reaches a maximum elevation of approximately 700 m above sea level (Mollah et al. 2004). The forest is broadly classified as mixed tropical evergreen and semi-evergreen forest (Das 1990). The vegetation originally consisted of tall, mixed evergreen trees, which occurred primarily in deep valleys and shaded slopes dominated by unique Garjan (*Dipterocarpus* spp.) trees. However, currently, the hills are mainly denuded and dominated by sun grass (*Imperata cylindrica*), herbs, shrubs, and brush woods. Tropical evergreen forests are found in deep valleys where wet conditions exist with shade. Tropical semi-evergreen forests predominate on the hills and flatlands. Evergreen species are more frequent in the lower stories; the main upper story has a high proportion of deciduous species that lose their leaves during the dry season. However, some places have experienced good natural regrowth, particularly of ground flora and the middle story, because of favorable climatic and edaphic conditions, thereby enhancing the inherent conservation value of the forest.

The sanctuary harbors a great diversity of tropical semi-evergreen flora and fauna. The TWS contains 55 mammals, 286 birds, 56 reptiles, 13 amphibians, and 290 plant species (BFD 2006; Khan 2008). It is also the last habitat for the Asian elephant (*Elephas maximus*), whose population varies from 15 to 100 (IUCN 2004). Additionally, large Indian civets (*Viverra zibetha* and *Viverricula indica*) are found in the TWS, which is the only recognized site of these animals in Bangladesh (Alam et al. 2012). It is home to avifauna of many species, which are dependent on good undergrowth and forest cover. Various non-timber forest products (NTFPs), including medicinal plants, bamboo, canes, sun grass, fish, and wild animals, are obtained from the forests.

2.4.3 An Analysis of Deforestation in Teknaf Based on Landsat Images

The severity of Teknaf's deforestation was the strongest motivating factor for conducting this research project. Landsat satellite images were analyzed to assess the extent to which forest degradation had advanced over time. A Landsat image consists of 30 × 30 m cells. Each cell is assigned with values measured by multiple sensors. Landsat image cells within the study area were processed and categorized according to the values of the normalized difference vegetation index (NDVI). NDVI is an index of the density of green plants on a patch of land and is calculated based on the difference in values between the visible red and the near-infrared sensors. Its value may range between −1.0 and 1.0, with 1.0 indicating most active green. NDVI has been extensively applied in forest-related analyses based on remote sensing data (e.g., Justice et al. 1985). First, Landsat cells that have less than 0.4 NDVI were first eliminated from the analysis because these places are not probably covered with much vegetation. The remaining cells were classified by changes in NDVI during the dry season of a particular year. For this study, cells were clustered using the Iterative Self-Organizing Data Analysis Technique (ISODATA) based on a series of NDVI values obtained for every month from the beginning of the dry season (November) to its conclusion (March–April) during any 1 year. As a result, three classes of cells were identified, of which Class 3 cells, which entailed the highest NDVI value, were postulated to contain mature trees or forest.

However, ground truth data collected at 63 locations classified as Class 3 indicated that these cells did not necessarily contain mature trees. This is because the presence of grassy plants and shrubs could also result in a high NDVI value. Therefore, a second step in the analysis conducted on Class 3 cells entailed segregating these cells into those with trees and those with only a few trees. For this step, one satellite image for a year was selected for analysis. The timing of the image was as close as possible to the end of the dry season, when the majority of grassy plants were likely to have died, so that the NDVI value reflected the presence of perennial trees. Each cell belonging to Class 3 within this selected image was analyzed, focus-

Table 2.4 Changes in the number of cells of the five vegetation classes in percentage between 1989 and 2015

Class	1989 (%)	2015 (%)	Change (%)
5	24.5	8.8	−16.5
4	22.7	20.5	−2.2
3	18.9	19.5	+0.6
2	29.3	44.7	+15.4
1	4.6	6.6	+2.0

ing on three types of spectral characteristics: NDVI, Normalized Difference Water Index (NDWI), and Green-Red Vegetation Index (GRVI). This analysis yielded three new clusters (3, 4, and 5) produced out of the former Class 3, resulting in a total of five classes. The presence of trees within the locations of the three new classes was verified using high-resolution satellite images and the ground truth method, confirming that cells within Classes 4 and 5 had trees.

Table 2.4 presents percentage changes over a period of 25 years starting from the 1988–1989 season and ending with the 2014–2015 season. The table shows that 48% of the land around the TWS was forested in 1988–1989 if we consider Classes 4 and 5 as "forested area." However, in 2014–2015, forests within the same classes occupied only 29% of the total area, evidencing a 38% reduction of forested area. The findings of this assessment generally concur with those of other estimates of forest reduction. Satellite images taken by the Bangladesh Space Research and Remote Sensing Organization show that the forested area decreased by 39% between 1972 and 1990 (Rahman et al. 2013). Between 1989 and 2009, the forest cover in the TWS, which is a protected area, decreased by 46% (CEGIS 2011). Islam (2011) estimated that 10% of the major tree species in the TWS only occur in a few scattered places. However, during this same period, the herb/shrub/bush cover increased from 53.9% (1989) to 67.4% (2009).

2.4.4 Current Condition of the Forest

Various factors are involved in reducing forest resources. Portions of the natural forests of the TWS were converted to long-rotation plantations. Various human activities, including pilferage, in the recent past are responsible for the loss of stocking. In addition to being adversely affected by human activity, many of the trees suffered heavily from cyclones during the last three decades. Since the 1990s, the decline of the forest area as a result of human settlements and frequent cyclones is remarkable. Human settlements have increased in and around forest areas, and most of the people are directly or indirectly dependent on the forest for their subsistence. In addition to clearing the forest for housing and agricultural activities, people use forest products for agricultural applications such as building shading structures for betel leaf cultivation. The BFD (2006) identified 29 stakeholders who are directly involved in forest resource extraction activities with major/moderate stakes. The stakeholders have been categorized into primary and secondary groups. Among the

22 primary stakeholders, fuelwood/timber collectors, betel leaf growers, forest produce collectors, hunters, and fishermen are the major ones, whereas among the seven secondary stakeholders, brick field owners, a timber/fuelwood merchant, a sawmill owner, and a tea stall owner have an indirect influence on forests.

The Forest Department started a social forest (SF) program along the border of the depleted area of the forests in 2003–2004. A total of 37 active SFs have been identified, of which 26 had less than 50% canopy cover in 2013 because of illegal tree felling, while seven SFs had been totally cleared (Asahiro et al. 2014).

2.4.5 Forest Policy and Management

The government approved the Forest Policy of 1994 by revising the first policy of 1979. It was developed based on an equitable benefit-sharing system with the local people, and it also enhanced people's participation in forest management activities. A 20-year Forestry Sector Master Plan (1993–2013) was approved in the 1990s, and its main objective was to bring 20% of the country's land under tree cover (Islam and Sato 2012). It was very difficult for the BFD to achieve a positive change in the country's forest condition through traditional forest management approaches. The trends of forest losses in Bangladesh have received enormous attention from policy makers, as well as donors, since Bangladesh gained independence in 1971. Thus, social forestry (which includes local people) approaches were officially introduced in 1981 with the support of the Asian Development Bank. In 2011, the Bangladesh Government approved Social Forestry Rules (an amendment) that have clearly outlined and described the different aspects of people-oriented forest management programs.

In Teknaf, in addition to the social forestry program, co-management activity has been adopted by the Forest Department. To address forest conservation and related economic opportunities, the Nishorgo Support Project (NSP) (2004–2008) was conducted in the TGR. A follow-up project entitled "Integrated Protected Area Co-management (IPAC)" was undertaken in collaboration with the Government of Bangladesh's Ministry of Environment and Forest and the Ministry of Fisheries and Livestock from 2008 to 2013. Under the NSP, various alternative income-generating activities (AIGAs) were distributed and found to be successful. Among the AIGAs, cow fattening, small trade businesses, nursery development, fish cultivation, homestead vegetable gardening, and improved stove (*chulha*) installation have been successful, while poultry rearing failed (Karim 2008). In a current investigation, Karim (2008) observed that AIGAs play a limited and inconsistent role in reducing forest dependence because of inadequate support and a lack of coordination, monitoring, and management.

2.5 People

Teknaf has a warm, tropical climate, and sufficient rainfall to support rich biological diversity, making it an attractive as well as an ecologically important place. Because of its geographical location, Teknaf is a site of political and historical importance. At various times, this area has been under the rule of Bengali sultans, Arakanese kings fleeing Afghans, the Portuguese, and the British (Egbert 1971). All of these regimes have influenced the language, culture, and social structure of the area. As a result, residents of Teknaf currently include Bengalis, Rohingya refugees, and small ethnic groups that share the same local resources and a similar culture. The people of Teknaf are predominantly poor and highly dependent on natural resources (the forest and sea). The main livelihoods in this area are rice and betel leaf cultivation, fuelwood collection from the forest, sea fishing, and daily labor work.

2.5.1 Approach

In addition to consulting previous scientific reports and survey records, we compiled data from a recent survey conducted in Teknaf upazila by the research team from Kyushu University with the help of local research assistants, which also profiled the population of the area. The survey included counting and collecting the locational information of all the households of Teknaf upazila (except the island under Teknaf upazila named St. Martin, because we considered only the mainland area for the study). It relied on village lists obtained from the Bangladesh Bureau of Statistics (BBS), which is the centralized official bureau responsible for collecting and disseminating demographic, economic, and other relevant statistics about the country. The present study also included direct interviews conducted with 10% of the households to obtain socioeconomic data. A systematic sampling approach was applied, selecting every tenth household to be interviewed. According to a report released by the BBS (2011), there are 153 villages, with a total population of nearly 46,000 households in Teknaf upazila. During our survey conducted from December 2015 to May 2016, we identified 147 villages comprising 57,404 households (Appendix I). Locational data on households were collected using Garmin GPSMAP 62 and 64 devices. Household heads were interviewed using a structured interview schedule containing questions on topics such as age, annual income, family size, education, and fuelwood use.

2.5.2 Demography

Eighty percent of the total population of Teknaf upazila resides in rural areas. The majority of the population comprises Muslims (0.25 million), followed by Buddhists, Hindus, and other religious groups. About 74% of the total population is aged below 30 years, indicating a high birth rate, with only 2.5% of the total population being over 65 years, which indicates a low life expectancy. The average rural household size is 6.2 persons, and that of urban households is slightly lower at 5.8. The literacy rate in Teknaf is 26.7%, whereas the national literacy rate is 57%. Of the total population in Teknaf, 9% has attained a primary education, 3% has attained a secondary education, and less than 2% has obtained a higher education. Approximately 98% of the population is Bengali, with less than 2% comprising other ethnic groups. In Teknaf there are 572 ethnic households comprising 3008 people (BBS 2011).

Apart from the local residents in Teknaf, there are Rohingya refugees (from Myanmar) living within a camp as well as in local settlements. Though they are not registered as legal residents of Bangladesh, each year they arrive in large numbers in the country. The exact number of refugees is unknown, but representatives of nongovernmental organizations have placed this figure at anything between 100,000 and 350,000 people (Sajjad 2003). According to our survey findings, there are 57,404 houses scattered across 147 villages. Though the survey did not cover the total population, based on our household interviews, we can predict that considering an average family size of 6.1 members, the population is around 0.35 million. Figure 2.12 shows the distribution of households across the villages. The villages in Teknaf have no official boundaries, so the demarcation was done on the basis of households. Table 2.5 shows the number of households per union (a union is an administrative unit under an upazila). Teknaf upazila comprises six unions, namely, Baharchhara, Nhilla, Whykhong, Sabrang, Teknaf Sadar, and St. Martin. For this study, we only considered unions located on the mainland (St. Martin is an island in the Bay of Bengal). We also surveyed villages under the Teknaf municipality. Table 2.4 presents a socioeconomic profile of the population of Teknaf. Of the surveyed unions, Baharchhara evidenced the highest annual incomes and the largest areas of land owned by households.

2.5.3 Livelihoods

Livelihoods comprise capabilities, assets, and activities required for subsistence (Chambers and Conway 1992). In many developing countries in the world, forest-related activities, in addition to being a main income source, are an integral part of the livelihoods of millions of people (Byron and Arnold 1999). When livelihood options are limited, economic activities may place forests under threat (Ghimere 1996), as poor people living in or on the periphery of natural forests or woodlands have easy access to forests (Calibre 2000). This situation applies to reserve forests

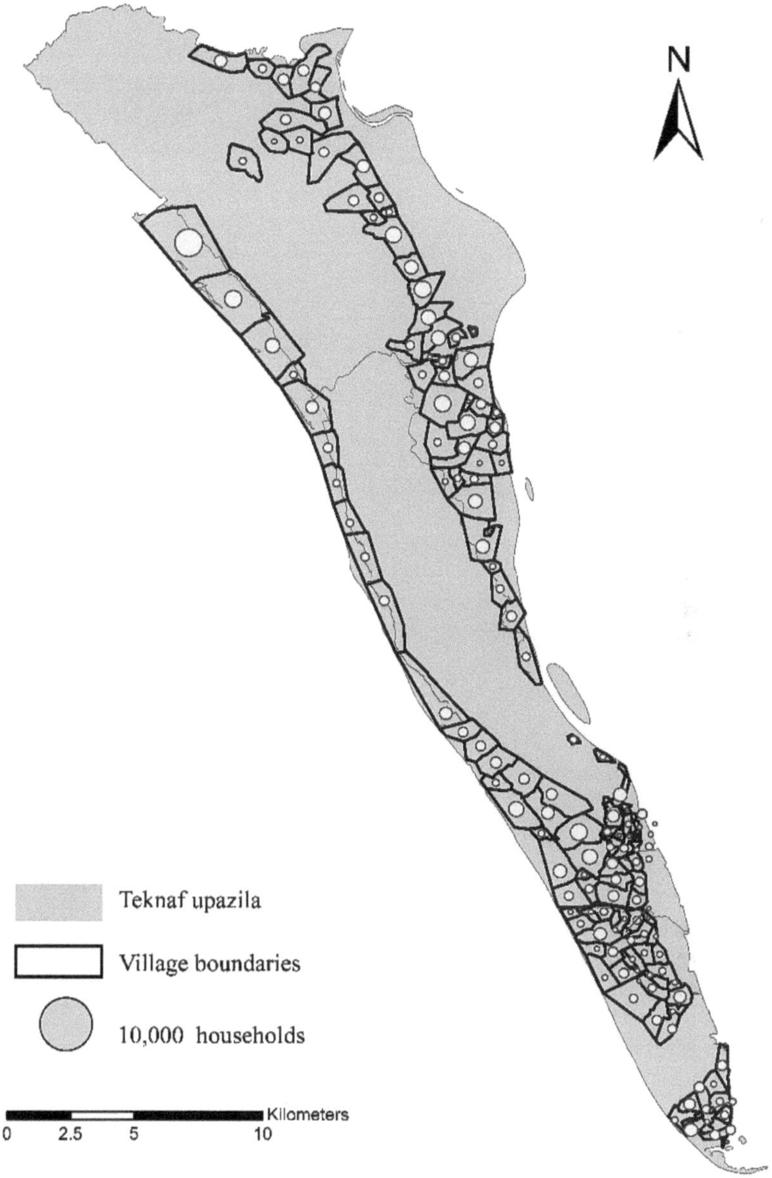

Fig. 2.12 Distribution of households in Teknaf

in Teknaf where local people's livelihoods depend exclusively on forests. Forest-related activities are an integral part of the livelihood activities and strategies of the extensive population in Teknaf because of poverty and easy forest access. Thirty-eight percent of Teknaf's population is poor and living below the poverty line (on

Table 2.5 Union-wise socioeconomic status of Teknaf

	Baharchhara	Nhilla	Whykhong	Teknaf Sadar	Teknaf town	Sabrang	Teknaf upazila
Number of houses	8674	10,605	12,185	11,010	3622	11,308	57,404
Mean duration of settlement (year)	21	25	19	31	28	28	26
Mean age of household head (year)	40	41	41	40	40	39	40
Mean family size (2016)	6.5	6.1	6.0	6.1	5.9	6.2	6.1
Mean education (schooling year)	1.7	2.5	2.2	2.0	2.0	1.2	1.9
Average annual income (000 BDT)	236	206	200	170	153	186	195
Mean land property (decimal of ha)	49.43	25.44	31.69	26.98	5.7	20.19	28.29

Table 2.6 Union-wise household livelihood activities

Number of households (Avg. income in 000 BDT)	Baharchhara	Nhilla	Whykhong	Teknaf Sadar	Teknaf pourashava	Sabrang	Teknaf upazila
Farming (avg. income 119)	52	129	218	53	0	68	520
Betel leaf farming (avg. income 130)	163	2	1	126	1	81	374
Fishing (avg. income 138)	154	48	30	153	10	231	626
Abroad (avg. income 292)	110	127	126	100	32	207	702
Business (avg. income 174)	160	216	202	211	127	200	1116
Others (avg. income 154)	89	285	201	235	118	171	1117

Table 2.7 Livelihood activities in Teknaf

No.	Livelihood activity	Overall household	Local people	Rohingya refugee
1	Fuelwood collection	52	35	87
2	Agro farming	41	55	17
3	Sun grass collection	34	27	87
4	Day labor	30	23	45
5	Salt production	30	46	
6	Fishing	23	25	16
7	Small business	22	25	13
8	Shrimp fry catching	18	17	20
9	Illicit felling	18	20	15
10	Bamboo and cane extraction	17	15	22
11	Fruits and vegetables	16	12	25
12	House building material collection	14	13	17
13	Grazing and fodder collection	9	15	
14	Medicinal plant collection	9	6	17
15	Green and dry leaf collection	7	4	13
16	Grocer	7	10	5
17	Betel leaf cultivation	5	8	
18	Hunting	5	2	12
19	Honey collection	3	2	5
20	Rickshaw pulling	3		8
21	Brick field owner	3		

US$1.25 per day). Poverty in this area is associated with low agricultural wage rates, which are in the range of 119–140 BDT (BBS 2011).

A finding of our survey, which also covered livelihood options, was that household members are engaged in several livelihood activities. However, we asked them to specify their main livelihood activities and household annual incomes. Table 2.6 shows numbers of households within each union engaged in different livelihood activities. The five main livelihood options in Teknaf were found to be farming, *pan* farming (*pan* is the local term for betel leaf), fishing, employment abroad, and business.

Other livelihood options included services, fuelwood collection, and jobs like teaching and salesmen. Table 2.6 shows that incomes derived from employment abroad and business are higher than those derived from natural resource-based livelihoods such as farming and fishing. Local people, including those from different ethnic groups and Rohingya refugees, are engaged to different extents and in different ways in various livelihood activities such as farming, fishing, labor, business, and extraction of forest products that impact on forests to greater or lesser degrees. Table 2.7 lists 21 different livelihood activities, identified by Uddin and Khan (2007) that local people and Rohingya refugees in Teknaf peninsula are engaged in.

2.5.4 Settlements

From the beginning of human history, people have lived in or near forests, because they are a source of food, shelter, and fuelwood. With its concentration of settlements within a narrow strip of land along the Bay of Bengal, Teknaf evidences this kind of forest-dependent living pattern. A total of 147 villages have been identified, of which 102 villages are located in or near reserve forests in Teknaf upazila (see Fig. 2.13). Apart from the area's local inhabitants, the population comprises ethnic communities of Chakma who moved to the peninsula during the mid-1800s (Rahman 2011), as well as another ethnic community, the Rakhine, who have also been living there for a long time. Eleven of the villages, namely, Amlatali, Chowdhuripara, Dakshin Shilkhali, Domdomia, Harikhola, Khrangkhali, Lambaghona, Naikhangkhali, Nhilla Bazar, Nhilla Mogpara, and Raikheong, comprise ethnic communities, mostly Rakhine and Chakma. Rohingya refugees generally live in the unions of Jahajpura, Shamlapur, and Teknaf, with only two of their settlements (Noyapara camps 1 and 2) having official legal status.

The current system of reserve forests was established through the Forest Act introduced in 1927 by the British colonial government prior to this time, and people living in the forests had legal status. The government subsequently gave households permission to live on 2 ha homestead plots in the forest and designated them as "for-

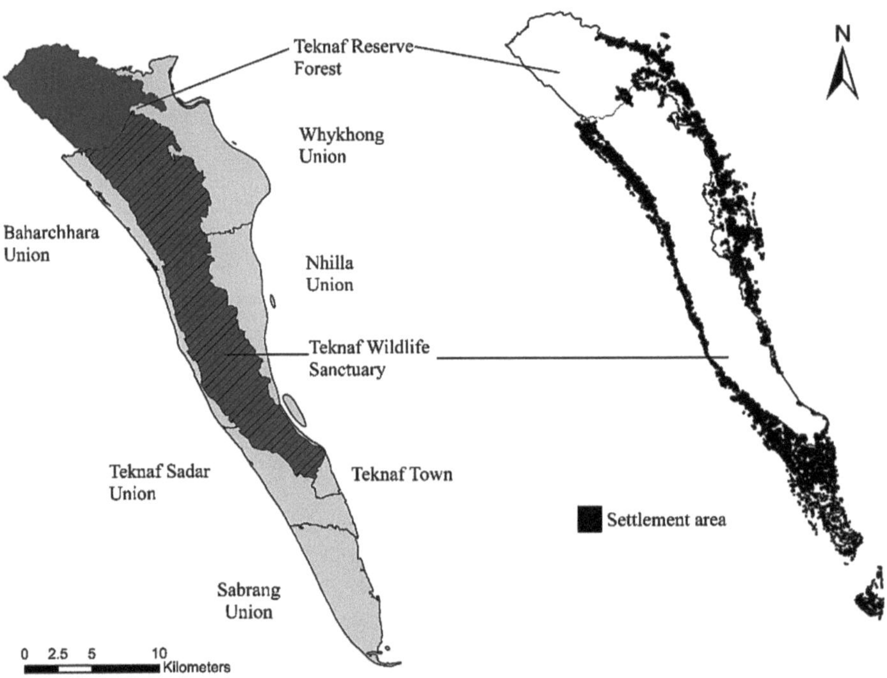

Fig. 2.13 Population settlement in Teknaf upazila around the forest

est villagers." The descendants of forest villagers also acquired the right to live in forests. Land use was minimal until the late 1980s when population pressure began to be felt. In the late 1970s, with the influx of refugees from Myanmar, and again in 1991, population expansion resulted in the large-scale conversion of forests as well as agricultural land for human settlement purposes (BCAS 1997). In Teknaf, with the increase in population, a number of households have acquired more land encroached from the forests. Some people, who lost their land and houses because of the construction of the Marine Drive, began building houses on hilltops, subsisting on forest resources. In 1990, following a major cyclone, many more people moved into forested areas from the coastal plains near the seaside. Presently, settlements are scattered and have begun to spread out within the forest. Settlement boundaries, starting from the beach or river side, extend all the way into the reserve forests. Thus, in addition to legal households, there are many encroachers now living in the forests, as discussed in detail in Chap. 8.

Appendix 2.1 Villages in Teknaf upazila

Union	Village	Households
Teknaf town	Kaikkali Para	45
Teknaf town	Naittong Para	801
Teknaf town	Puran Pollan Para	622
Teknaf town	Kaikal Para	171
Teknaf town	Islamabad	235
Teknaf town	Oilabad	71
Teknaf town	Oilabad	240
Teknaf town	Dalipara	72
Teknaf town	Kulal Para	290
Teknaf town	Chowdhury Para	93
Teknaf town	Uttar Jalia Para	386
Teknaf town	Bazar Para	63
Teknaf town	Madhya Jalia Para	121
Teknaf town	Dakhin Jalia Para	229
Teknaf town	Hungar Para	22
Teknaf town	Kulal Para	161
Baharchhara	Hajam Para	376
Baharchhara	Mathabhanga	227
Baharchhara	Bara Dail	309
Baharchhara	Kachapia	408
Baharchhara	Noakhali	684
Baharchhara	Marish Bania	252
Baharchhara	Uttar Shilkhali	1276
Baharchhara	Shamlapur	3320
Baharchhara	Jahajpura	678
Baharchhara	Halbania	237
Baharchhara	Dakshin Shilkhali	907
Nhilla	Naikhangkhali	891
Nhilla	Hoabrang	354
Nhilla	Pankhali	1186
Nhilla	Sikdar Para	1032
Nhilla	Lechuaprang	239
Nhilla	Ulochamari	675
Nhilla	Nhilla Mogpara	405
Nhilla	Nhilla Bazar (Bazar Para)	209
Nhilla	Fullerdail	542
Nhilla	Nath Murapara	302
Nhilla	Kona Para	107

Union	Village	Households
Nhilla	Chowdhury Para	111
Nhilla	Purba Rangikhali	276
Nhilla	Paschim Rangikhali	214
Nhilla	Jumma Para	150
Nhilla	Alikhali	826
Nhilla	Puchinga Para	48
Nhilla	Leda Para	802
Nhilla	Muchani Para	170
Nhilla	Nayapara	270
Nhilla	Jadimura	461
Nhilla	Dumdumia	281
Nhilla	Huakya Para	58
Nhilla	Marichaghona	229
Nhilla	Rojarghona	263
Nhilla	Ali Akbar Para	504
Sabrang	Baharchhara	110
Sabrang	Chandauli Para	230
Sabrang	Uttar Nayapara	97
Sabrang	Kyurabuja Para	196
Sabrang	Fathe Ali Para	94
Sabrang	Hariakhali	342
Sabrang	Hadurchhara	108
Sabrang	Mundar Dail	667
Sabrang	Mondal Para	264
Sabrang	Sikdar Para	243
Sabrang	Mogpara	49
Sabrang	Panchhari Para	207
Sabrang	Benga Para	113
Sabrang	Acharbania	150
Sabrang	Lezir Para	89
Sabrang	Koanchhari Para	426
Sabrang	Deguliar Bil	231
Sabrang	Dail Para	136
Sabrang	Dakshin Nayapara	627
Sabrang	Puran Para	309
Sabrang	Katabania	310
Sabrang	Kachubania	397
Sabrang	Lafarghona	117
Sabrang	Zinnah Para	127
Sabrang	Pendal Para	169
Sabrang	Alirdeil Para	387
Sabrang	Kurer Mukh	133
Sabrang	Guchha Gram	100
Sabrang	Rullher Depa	143

Union	Village	Households
Sabrang	Karachi Para	70
Sabrang	Khairtipara	69
Sabrang	Mistry Para	410
Sabrang	Purba Uttar Para	243
Sabrang	Golapara	40
Sabrang	Dakshinpara	441
Sabrang	Hajir Para	107
Sabrang	Paschim Uttar Para	464
Sabrang	Bazar Para	290
Sabrang	Majher Para	773
Sabrang	Jalia Para	441
Sabrang	Majer Dail	165
Sabrang	Dangor Para	561
Sabrang	Dail Para	188
Sabrang	Karachi Para	180
Teknaf Sadar	Razarchhara	420
Teknaf Sadar	Habibchhara	424
Teknaf Sadar	Mitta Panirchhara	539
Teknaf Sadar	Dargachhara	203
Teknaf Sadar	Tulatali	195
Teknaf Sadar	Lambori	980
Teknaf Sadar	Hatiarghona	467
Teknaf Sadar	Lengurbil	726
Teknaf Sadar	Jahalia Para	540
Teknaf Sadar	Kerantali	215
Teknaf Sadar	Baraitali	152
Teknaf Sadar	Natun Pallan Para	1195
Teknaf Sadar	Shilbania Para(Part)	128
Teknaf Sadar	Goder Bil	972
Teknaf Sadar	Dail Para (Part)	491
Teknaf Sadar	Mohish Khalia Para	784

References

AIS (2017). Agriculture information service. Ministry of Agriculture, Government of the People's Republic of Bangladesh, Dhaka

Alam MM, Hossain MA, Shafee S (2003) Frequency of Bay of Bengal cyclonic storms and depressions crossing different coastal zones. Int J Climatol 23:1119–1125

Alam MF, Uddin MZ, Hasan MA (2012) Evaluation of plant biodiversity in Teknaf wildlife sanctuary, Bangladesh. LAP LAMBERT Academic Publishing, Saarbrücken

Asahiro K, Nakagaki A, Rahman MA, Tani M (2014) Current conditions of social forests in Teknaf and Shilkhali ranges of Bangladesh. In: Proceedings of the international conference on environmental aspects of Bangladesh (ICEAB 2014), University of Dhaka, Bangladesh

Bari A, Dutta U (2004) Co-management of tropical forest resources in Bangladesh. Secondary data for pilot protected area: Teknaf game reserve, USAID-Bangladesh and Ministry of Environment and Forest, Govt. of People's Republic of Bangladesh

BBS (2011) Preliminary report of census. Bangladesh Bureau of Statistics, Ministry of Planning, Government of the People's Republic of Bangladesh, Dhaka

BBS (2015) Statistical year book. Bangladesh Bureau of Statistics, Ministry of planning, Government of the People's Republic of Bangladesh, Dhaka

BCAS (1997) Bangladesh center for advance studies – biological survey; Ministry of Forest Environment and Forest, Govt. of People's Republic of Bangladesh

Belal MAH (2013) Effectiveness of co-management committees in Teknaf Wildlife Sanctuary. In: Fox J, Mustafa MG, Bushley Brennan SM, Durand L (eds) Connecting communities and conservation: co-management initiatives implemented by IPAC in wetland and forests of Bangladesh. USAID-Nishorgo Network, Dhaka

BFD (2006) Management plans for Teknaf game reserve, Nishorgo supported project, Bangladesh Forest Department. Ministry of Environment and Forest, Govt. of People's Republic of Bangladesh

BFD (2014) Wildlife sanctuary: protected areas of Bangladesh. Bangladesh Forest Department. [Updated 2011 Dec 15; cited 2014 December 03]. Available from: http://www.bforest.gov.bd/index.php/protected-areas

BMD (2014) Weather report, Bangladesh Meteorological Department. Government of the People's Republic of Bangladesh

Byron N, Arnold M (1999) What features for the people of the tropical forest? World Dev 27(5):789–805

Calibre SSC (2000) Number of forest-dependent people: a feasibility study for DFID's forestry research programme. Calibre Consultant and the Statistical service Center, University of Reading, Reading, pp 1–84

CEGIS (2011) Landuse change trend analysis in seven protected areas in Bangladesh under IPAC through application of landsat imageries. USAID's Integrated Protected Area Co-Management (IPAC) project. USAID/Bangladesh

Chambers R, Conway GR (1992) Sustainable rural livelihoods: practical concepts for the 21st century, Discussion paper 296. Institute of Development Studies, Brighton

Das DK (1990) Forest types of Bangladesh, Bulletin No. 6, Plant taxonomy series. Bangladesh Forest Research Institute, Chittagong

Dasgupta S, Huq M, Khan ZH, Ahmed MMZ, Mukherjee N, Khan MF, Pandey K (2010) Vulnerability of Bangladesh to cyclones in a changing climate – potential damages and adaptation cost. Policy Research Working Paper. The World Bank Development Research Group, Environment and Energy Team

Edwards DC, McKee TB (1997) Characteristics of 20th century drought in the United States at multiple timescales, Climatology Report No. 92–7. Colorado State University, Fort Collins

Egbert S (1971) Arakan and Bengal: the rise and decline of the Mrauk U Kingdom (Burma) from the fifteenth to the seventeenth century AD

FAO-UNDP (1998) Land resources appraisal of Bangladesh for agricultural development. Report no. 2, Agro-ecolgical regions of Bangladesh. BGD/81/035, tech. Report no. 2. FAO, Rome, 570 p

Ghimere K (1996) Refugees and deforestation. In: Birendra KC, Shin N (eds) Refugee impact on collective management of forest resources: a case study of Bhutanese refugees in Nepal's Eastern Terai region. J For Res 11:305–311

GoB (2009a) Bangladesh climate change strategy and action plan 2009. Government of the People's Republic of Bangladesh

GoB (2009b) Government of Bangladesh gazette notification no. moef/forest-2/02/wildlife/15/2009/492 dt.09/12/2009. Ministry of Environment and Forest, Govt. of People's Republic of Bangladesh

Green MJB (1987) World conservation monitoring centre, IUCN commission on natural parks and protected areas. Teknaf game reserve, IUCN Directory of South Asian Protected Areas, pp 38–43

Hayes M, Svoboda M, Wilhite D, Vanyarkho O (1999) Monitoring the 1996 drought using the SPI. Bull Am Meteorol Soc 80:429–438

Islam MS (2011) Identification and assessment of threats of forest conservation in Teknaf Wildlife Sanctuary. MS thesis, Institute of Forestry and Environmental Sciences, Chittagong University, Bangladesh

Islam KK, Sato N (2012) Deforestation, land conversion and illegal logging in Bangladesh: the case of the Sal forests. iForest 5:171–178

IUCN (2004) Conservation of Asian elephants in Bangladesh. The World Conservation Union, Bangladesh

Justice CO, Townshend JRG, Holben BN, Tucker CJ (1985) Analysis of the phenology of global vegetation using meteorological satellite data. Int J Remote Sens 6:1271–1318

Karim MN (2008) Local knowledge of indicator birds: implications for community-based ecological monitoring in Teknaf game reserve. In: Fox J et al (eds) Connecting communities and conservation: collaborative management of protected areas in Bangladesh. East-West Center, Honolulu, pp 139–160

Khan MMH (2008) Protected areas of Bangladesh- a guide to wildlife. Nisorgo program, Bangladesh Forest Department, Dhaka, Bangladesh

McKee TB, Doesken NJ, Kleist J (1993) The relationship of drought frequency and duration to time scales. In: Proceeding of the 8th conference on applied climatology. American Meteorological Society, Boston, pp 179–184

Mollah AR, Rahman MM, Rahman MS (2004) Site-level field appraisal for protected area co-management: Teknaf game reserve. Dhaka, Nature Conservation Management (NACOM)

Pender JS (2010) Climate change, its impacts and possible community-based responses in Bangladesh, 2nd ed. Church of Bangladesh, Dhaka

Rahman MA (2011) Deforestation and forest conservation in a Tanchangya community rural livelihoods and protected landscapes: co-management in the Wetlands and Forests of Bangladesh, 120

Rahman MZ, Tani M, Moslehuddin AZM, Ullah SMA (2013) Use of information sources in maintaining livelihoods by Rohingya refugees around Teknaf wildlife sanctuary. J Agrofor Environ 7(2):15–18

Rahman MA, Tani M, Asahiro K, Rahman MZ, Moslehuddin AZM (2014) Impacts of climate change and land use on forest degradation in Teknaf Peninsula. Int J Environ 4(2):46–51

Rosario EA (1997) The conservation management plan of the protected areas other than those in Sundarban forest of Bangladesh. GOB/WB forest resource management project in Bangladesh

Sajjad T (2003) SRI on-site action alert: Rohingya refugees of Burma and UNHCR's repatriation program. Asia Researcher, Survivors' Rights International, Annapolis

SRDI (2001) Land and soil resources utilization guide (in Bengali). Upazila Nirdeshika series, Teknaf upazila, Cox's Bazaar district. Soil Resource Development Institute, Dhaka, 166 p

UAO-Teknaf (2016) Report of Upazila Agriculture Officer (UAO), Teknaf, Department of Agricultural Extension, Ministry of Agriculture, People's Republic of Bangladesh

Uddin MS, Khan MASA (2007) Comparing the impacts of local people and Rohingya refugees on Teknaf Game Reserve, Bangladesh. In: Fox J, Bushley B, Dutt S, Quazi SA (eds) Making conservation work: linking rural livelihoods & protected area management in Bangladesh. East-West Center and Nishorgo Program of the Bangladesh Forest Department, Hawaii/Dhaka, pp 149–175

Williams M (2002) Climate change information kit. The United Nations Environment Programme and the Climate Change Secretariat (UNFCCC) and sponsored by UNEP, the UN Development Programme, the UN Department of Economic and Social Affairs, the UN Institute for training and research, the World Meteorological Organization, the World Health Organization, and the UNFCCC, Geneva, Switzerland

Wu H, Svoboda M, Hayes M, Wilhite D, Wen F (2007) Appropriate application of the standardized precipitation index in arid locations and dry seasons. Int J Climatol 27:65–79

Part II
Causes of Deforestation

Chapter 3
Impact of Betel Leaf Cultivation on Forests

Masakazu Tani

Abstract The betel is a major cash crop on the Teknaf Peninsula. The cultivation of betel plants affects forests in two ways. The construction of "*pan boroj*," a facility to provide the plants with shading, consumes a large amount of forest resources every year. Another way of betel cultivation's influence on forests is tree clearance in cultivation plots. *Pan boroj* tend to be constructed on gentle slopes where forests originally existed because more productive and limited flatlands are used for other crops such as rice. Trees are cleared to make room for a pan boroj, and as long as betel is cultivated, trees will not regrow at that spot. This study conducted village household surveys to record all pan boroj in a village and to quantitatively assess the impact of betel cultivation on the nearby forests. An estimate of forest resource use is equivalent to approximately 5% of biomass annually generated in the village area. Cultivation plots of betel plans also account for another 5% of the area. Despite the negative impacts of betel cultivation on forests, this study also finds the economic importance of betel cultivation in the village economy, and estimates that betel cultivation may compensate the lack of sufficient livelihood generated by rice cultivation in limited flat land.

Keywords Betel leaf cultivation • Rice farming • Forest resources • Forest degradation • *Pan boroj*

3.1 Introduction

The betel (*Piper betle*) is the leaf of a vine belonging to the *Piperaceae* family. The betel plant is an evergreen and perennial creeper, with glossy, heart-shaped leaves and white catkins. The betel plant originated in South and Southeast Asia (Varier 1995). Betel leaves are consumed as "*pan*", a stimulant and psychoactive quid in which pieces of areca nuts are wrapped in a leaf with slaked lime paste. Betel leaf chewing is widely prevalent in South and Southeast Asia (Rooney 1993).

M. Tani (✉)
Department of Environmental Design, Faculty of Design, Kyushu University, Fukuoka, Japan
e-mail: tani@design.kyushu-u.ac.jp

M. Tani, M.A. Rahman (eds.), *Deforestation in the Teknaf Peninsula of Bangladesh*, DOI 10.1007/978-981-10-5475-4_3

Approximately 10–20% of the world's population is estimated to engage in betel chewing (Gupta and Warnakulasuriya 2002). In Bangladesh, it is estimated that 30% of adults chew betel quid, and the rate is higher in rural areas (43%) than urban areas (19%) (Flora et al. 2012), which creates a large demand for betel leaves. Bangladesh is one of the largest consumers of betel leaves, and it produces approximately 100,000 tonnes of betel leaves. According to the Export-Import Databank of India, the largest destination for Indian betel leaves is Bangladesh, accounting for 80% of the betel leaves exported from India.

The betel is a major cash crop on the Teknaf peninsula, as it is in many other areas (Ghosh and Maiti 2011), and leaves are brought to larger markets in Bangladesh, such as Cox's Bazar and Dhaka, whereas the betel nut is cultivated throughout Bangladesh. Southeast Bangladesh, including the Teknaf peninsula, is famous for producing high-quality betel leaves, including "Maheshkhali *pan*" (Maheshkhali is the name of the region that is located north of the peninsula in the same district). To produce high-quality leaves, saplings are renewed every year, and the plants are cultivated in a facility called "*pan boroj*" to provide appropriate shading. As such, this system of cultivation that uses shading structures is called the *boroj* (conservatory) system (Kumar 1999), although in northern parts of Bangladesh, the betel is also cultivated with betel nut trees in agroforestry projects in open environments.

The size of a *pan boroj* varies, but a typical *boroj* may be approximately 20 m by 20 m (0.04 ha). *Pan boroj* are made of wood, bamboo, and leaves, which are renewed almost every year. These building materials mainly come from the forest in Teknaf. Because of the high density of *pan boroj* in this area, and because each *pan boroj* consumes a certain amount of forest products, betel leaf cultivation appears to be one of the major causes of deforestation in this area.

The objective of this chapter is to assess the impact of betel leaf cultivation on deforestation, based on detailed surveys of a village located along the west coast of the peninsula. This chapter first examines the status of betel leaf cultivation within the entire range of subsistence activities of the village to characterize the status of betel leaf cultivation in the local society. Two impacts of betel leaf cultivation on the forest are analyzed: the use of forest resources to construct *pan boroj*, and clearing forest areas to construct *pan boroj*. The first type of impact depletes forest components by exploiting trees, bamboo, and leaves. Regarding the second impact, an area of the forest is cleared to build a *boroj*, and as long as the *boroj* is in operation, trees will not regrow in the area.

3.2 Methodology

3.2.1 Study Area

The study area was the village of Marishbunia, which is located along the west coast of the peninsula approximately 30 km northwest of the Teknaf upazila (Fig. 3.1). Many small settlements are scattered all along the coast, and narrow strips of settlements are found in Marishbunia. The village is bordered by two minor streams, and

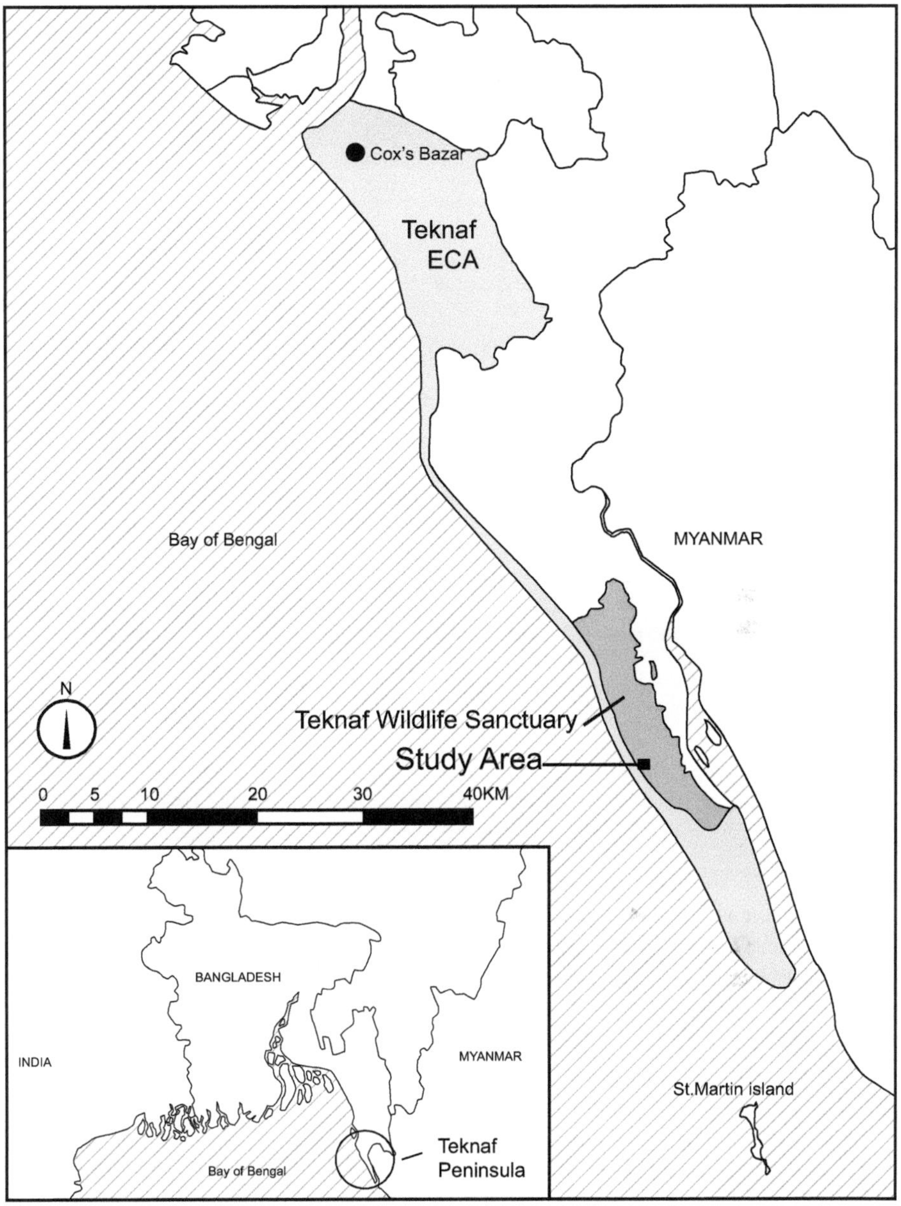

Fig. 3.1 Location of Marishbunia

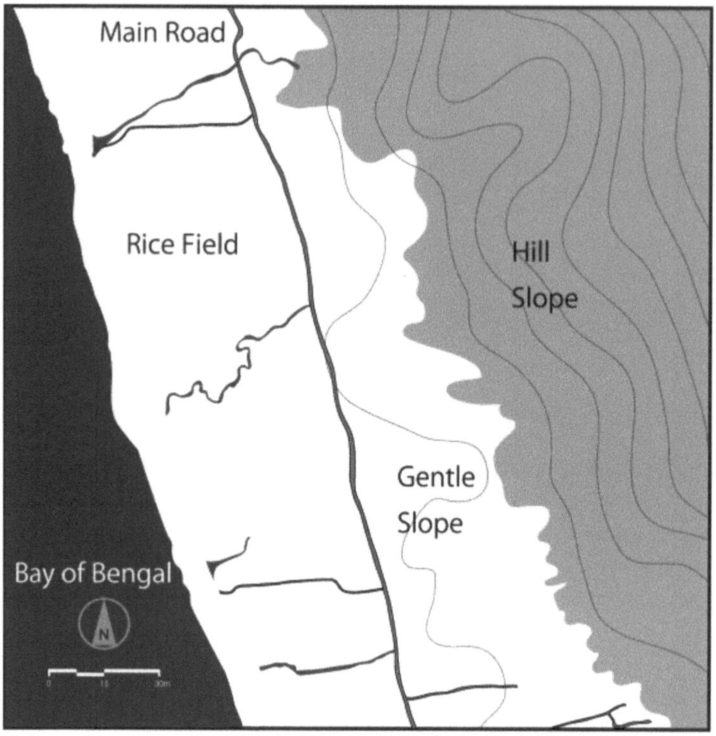

Fig. 3.2 Map of Marishbunia

its north–south span measures approximately 1.5 km (Figs. 3.2 and 3.3). Villagers recognize their village as extending from the beach to the hilltop behind their village, although there are no formal administrative village boundaries. In Marishbunia, the distance between the beach and the hilltop is also approximately 1.5 km. Therefore, the village lies in a roughly 2-km^2 area along the beach; a narrow (300-m-wide) strip of farmland occupies all of the flat area. From the end of the farmland, a gentle slope rises to the foot of the hill to the east. A single paved road, the main road, runs in the middle of the settlement parallel to the coast.

Until the 1970s, a rich forest existed on the hill and on the gentle slope up to the main road, but now there are only a few trees, shrubs, and vines on the hill, and houses and *pan boroj* have crept up the slope. An extensive area of the hill is currently utilized for sun grass (*Imperata cylindrica*) cultivation. Figure 3.3 illustrates a cross-section of the landform from the coast (west) to the hilltop (east), showing beach, rice field, gentle slope, and steeper hill slope areas. Marishbunia has 207 households and 1294 people according to our survey in 2010. Most houses are located on the gentle slope. There used to be houses along the coastal low-lying areas, but they were washed away by strong cyclones in the 1990s. People did not reconstruct these houses, but relocated up the slope.

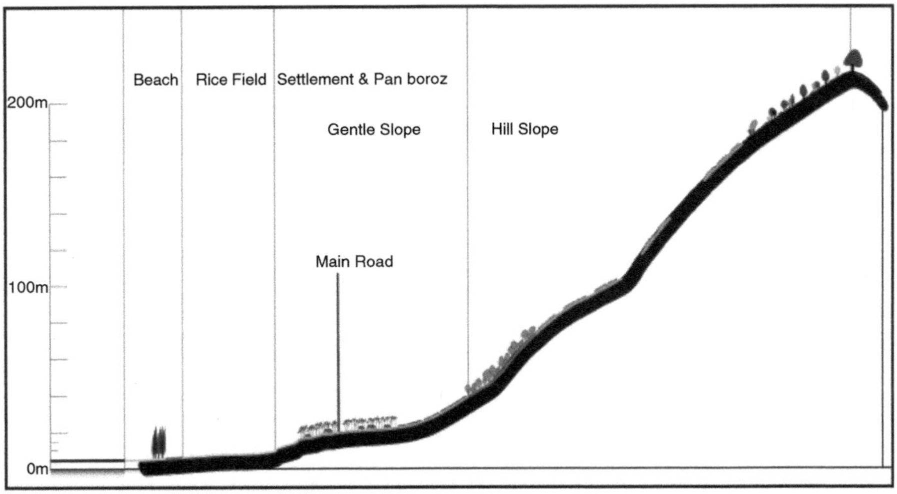

Fig. 3.3 Cross-sectional map of Marishbunia

3.2.2 *Survey and Analytic Method*

Surveys were conducted three times during 2010 and 2011 in Marishbunia. These surveys were funded by a grant from The Sumitomo Foundation for Environmental Studies, and were conducted jointly by Bangladesh Agricultural University, Bangabandhu Sheikh Mujibur Rahman Agricultural University, and Kyushu University.

In cooperation with the Coastal Wetland Biodiversity Management Project, the first survey (March 2010) visited several settlements along the west coast of the Teknaf peninsula to select a study area. Marishbunia was selected for this study because it has an active village organization, and its compact size enabled all of its households to be surveyed in one season. Additionally, its residents agreed to take part in the survey.

In the second season (September 2010), a team surveyed all of the households in the village via a questionnaire, and recorded the land use and vegetation around the village. For the household survey, members of the research team visited each household in the village with local assistants. The questionnaire used in this survey was written in English, but interviews were conducted in Bengali, with the aid of the local assistants. The interview items in the questionnaire addressed household composition, occupation, land tenure, subsistence/livelihood activities, fuelwood- and sun grass-gathering activities, and environmental awareness.

In the third season (March 2011), a survey specifically focused on betel leaf cultivation. It covered all individual *pan boroj*, their locations, the farmers involved, the status of land (owned or rented), *boroj* size, and the duration of cultivation.

Fig. 3.4 *Pan boroj* immediately after planting. The roof and walls are yet to be constructed

Pan boroj locations were measured using the Global Positioning System, but the size of a *boroj* was reported by the farmer, rather than by actual measurements. To quantitatively assess the amount of forest resources consumed by a *pan boroj*, all of the components and materials of one *pan boroj* were recorded. The amount of materials used in a *pan boroj* vary depending on the size of the shading structure. However, because the structure of each *pan boroj* is similar, the total amount of materials used by all of the *pan boroj* in the village can be estimated using the number of *pan boroj* and the area of each *boroj*. The size of *pan boroj* varies from about 200 m^2 to 800 m^2. They have a flat box shape with a height of approximately 2 m. Figure 3.4 shows a *pan boroj* immediately after planting, with the roof and walls yet to be constructed. Figure 3.5 shows a completed *pan boroj* with roof and walls for shade. The inside of a *pan boroj* is shown in Fig. 3.6. Figure 3.7 shows the structural poles used in a *pan boroj*.

3.3 Results and Discussion

3.3.1 Comparison of Rice and Betel Leaf Farming

The main industry of this area is agriculture. The total area of farmland in the village is 23.8 ha, or 0.11 ha per household. This is less than one-quarter of the national average (0.46 ha) of farmland holdings per farming household. Among the 207

Fig. 3.5 Completed *pan boroj* with roof and walls for shade

households in the village, only 43 (21%) own farmland, and 164 are landless. The average size of the farmland of these 43 households is 0.55 ha, while the national average size of farmland per landed household is 0.51 ha.

Although we did not record the actual farming yield, the yield of rice was estimated based on the size of the rice fields. According to the villagers, the production of unhusked rice in a field that is one *kani* (0.16 ha) is approximately 10 mounds (373 kg). Assuming that all of the farm plots in the village are rice fields, the production of unhusked rice would be 55.5 tonnes. This was converted into 268 kg per household (of all categories), or 1.3 tonnes per landed household. Studies show that the amount of rice consumed per capita in Bangladesh ranges from approximately 300 to 750 g of husked rice (Bae et al. 2002; Choudhury et al. 2001; Smith et al. 2006; Tani et al. 2012; Watanabe et al. 2004). Assuming the amount of rice consumption is 500 g per day per capita, yearly rice consumption would be 180 kg of husked rice, or 260 kg of unhusked rice. The average household size of this village is 6.3 persons; thus, the average household would require 1.6 tonnes of unhusked rice per year. This suggests that the farmland productivity in this village is so low

Fig. 3.6 Interior of a *pan boroj* showing the planting geometry

that even households that own farmlands may not be self-sufficient (Box 3.1). These results show that the farmlands of this village are too small to provide villagers with enough rice, and that many households may have to purchase rice to survive. Therefore, they need income-generating activities for their livelihood.

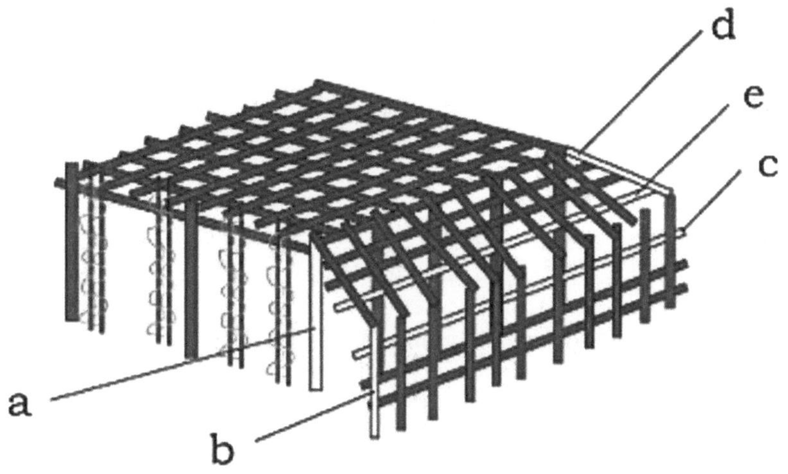

Fig. 3.7 Structural poles (*a–e*) used in a *pan boroj*

Box 3.1 Calculation of the Balance Between the Production and Consumption of Rice in Marishbunia Village

Production:

Production of unhusked rice = 10 mounds* (373 kg) per *kani*** (0.16 ha), or 2331 kg per ha.

Production of unhusked rice of the whole village: 2331 kg × 23.8 ha = 55,477.8 kg = 55.5 tonnes.

Production of unhusked rice per household (all): 55.5 tonnes ÷ 207 = 0.268 tonnes = 268 kg.

Production of unhusked rice per landed household: 55.5 ÷ 43 = 1.3 tonnes 1300 kg.

Consumption:

Average rice consumption per capita: 0.5 kg per day × 365 days = 182.5 kg

Rice consumption per household (6.3 persons): 182.5 kg × 6.3 = 1150 kg (husked rice)

Annual consumption of unhusked rice: Assuming the ratio of husked and unhusked rice is 0.7, then 1150 kg ÷ 0.7 = 1643 kg

*mound: the unit of weight, 1 mound equals 37.3 kg; ** *kani*: the unit of land, 1 *kani* = 0.16 ha

Table 3.1 Summary of ownership of farmlands (excluding *pan boroj*) and *pan boroj*

	Landless households	Landed households	Total
Farmland	164	43	207
Area (ha)	Total farmland	Farmland per household	Farmland per landed household
	23.8	0.11	0.55
No. of *pan boroj*	Total area (ha)	Average area (ha)	No. of *pan* farming households
152	9.3	0.062	115

While only 43 households own farmland, 115 households engage in betel leaf cultivation (Table 3.1). The total area occupied by *pan boroj* is 9.3 ha (57.6 *kani*), and the average size of a *pan boroj* is 0.062 ha (0.4 *kani*). Because some households operate more than one *pan boroj*, the average cultivated area per household engaged in betel leaf cultivation is 0.081 ha or 0.5 *kani*. Therefore, the area of betel farmland per operating household is much smaller than that of other farmlands. One factor that accounts for the small farmland area is the highly labor-intensive nature of betel leaf cultivation; thus, the area of land for betel leaf cultivation per farming household is rather limited. According to a focus group discussion on betel leaf cultivation, the most labor-consuming activity in betel leaf cultivation is watering plants. Water must be applied at least twice per week. Without any facilities such as sprinklers, the labor involved in watering includes fetching water from a nearby source, transportation of the water by manual labor, and actual watering in a *pan boroj*. Along with this time-consuming and physically demanding work, betel plants require frequent applications of agrochemicals, such as fertilizers, pesticides, and fungicides, because they are susceptible to diseases and pests.

Despite these difficulties, the income generated by betel leaf cultivation can be substantial. According to farming households, a typical half-*kani* (0.04 ha) *pan boroj* may generate approximately 70,000 Bangladeshi Taka (BDT) per season (approximately 6 months), provided no great difficulties, such as pests and diseases, are encountered. Given that rice production is approximately 10 mounds per *kani*, the amount of rice produced in the same size field as a typical *pan boroj* (0.5 *kani*) would be approximately 200 kg. If this 200 kg rice were sold, it would only yield 5000 BDT per season (at the rate of 25 BDT per kg of rice). This shows that betel leaf cultivation is much more productive than rice farming in terms of cash generation. Box 3.2 presents a calculation of income for the condition in which rice is cultivated on all 23.8 ha of regular farmlands and betel leaves are cultivated in all 9.3 ha of the *pan boroj* in the village. The results suggest that the income from betel leaf cultivation potentially supplement the insufficient income generated from regular farmland. Because betel plants can be grown on slopes, they do not compete for space on precious flatlands with other crops, which is another advantage of betel leaf cultivation. Therefore, many *pan boroj* tend to be placed on low slopes where the forest once thrived.

Box 3.2 Potential Income from Rice Production on Regular Farmland and Betel Leaf Production in pan boroj

Rice sale:

Assuming that rice production per ha = 2331 kg (see Box 3.1) and the price of rice (per kg) = 25 BDT, then the income from rice is 23.8 ha × 2331 kg × 25 BDT = 1,386,945 BDT.

Betel leaf sale:

Assuming that the income from a 0.5-*kani pan boroj* = 70,000 BDT, then the income from betel leaves is 58.1 *kani* (the total area of all *pan boroj*) ÷ 0.5 *kani* × 70,000 BDT = 8,134,000 BDT.

Area of rice farmland needed to replace the income from betel cultivation:

8,134,000 BDT ÷ 25 BDT per kg = 325,360 kg of rice, which requires 139.6 ha of rice fields.

3.3.2 Impact of Betel Leaf Cultivation on the Forest

This section examines the amount of forest resources consumed by betel leaf cultivation in building and maintaining *pan boroj*. To estimate the amount and kinds of materials necessary to construct a *pan boroj*, an actual *pan boroj* was measured and recorded. This *pan boroj* occupies 590 m^2 (0.37 *kani*), measuring 14.2 m by 41.5 m. It used 1103 wooden poles that were 5 to 10 cm in diameter and 2–3 m in length to make its basic structure, as well as 1324 thinner poles of wood and bamboo. There were 8280 sticks that supported the vines, and 50 bundles of sun grass were used to cover the walls and roof to create shade for the plants inside the *pan boroj*. Because all *pan boroj* are built anew every year, most materials also need to be newly acquired. Although some structural poles may be reused, at least one-half of all poles would be new materials.

The average size of a *pan boroj* in the village is 0.38 ha, which is almost identical to that of the recorded one. Because there is little structural variation among different *pan boroj*, using the measured Fig. 3.7 as a guideline, the total forest resources consumed by all *boroj* in the village would include: 170,000 structural poles, 200,000 thinner poles, 1300,000 supporting sticks, and 8000 bundles of sun grass.

Because it is difficult to quantitatively assess the amounts of thinner poles, sticks, and grass, the following estimate of resource consumption by *pan boroj* incorporates only structural poles into the calculation. Usually, poles are tied into bundles, and each bundle contains about 20–30 poles and weighs 20–30 kg. Therefore, if one pole were assumed to weigh 1 kg, the total weight of all poles used in *pan boroj*

Table 3.2 Construction of *pan boroj* over time.

Construction year	Number of *boroj*	%	Cumulative number of *boroj*	%
1950s	1	1	1	1
1960s	6	4	7	5
1970s	11	7	18	12
1980s	29	19	47	31
1990s	37	24	84	55
2000s and after	68	45	152	100

would be 170 tonnes. Because it was not possible to identify a definite area from which the people of Marishbunia extract forest resources, a rough estimate of the forest area was used. Given that the area of the village is roughly 1.3 km by 1.5 km, approximately two-thirds of which consist of gentle slopes and hills with forests, the area used to extract forest resources is 130 ha. The productivity of forest ecosystems in the tropics, based on data in Thailand, is 28.6 tonnes per ha (Tsuruta et al. 2012). Using this figure of forest productivity, the 130 ha of forest in Marishbunia would produce 3718 tonnes of plant biomass annually. Thus, the 130 tonnes of structural poles for the *pan boroj* is equivalent to 4.6% of the annual forest production of the village, which is equivalent to the production of 6 ha of forest. Currently, the forest in the village has lost the capacity to provide all of the materials needed to construct the village's *pan boroj*. As a result, a large amount of material is being transported from the eastern side of the peninsula and sold in the village. Some villagers also expressed sentiments that *pan boroj* in the village would become unsustainable in the near future because of the depletion of forest resources.

3.3.3 Impact of *pan boroj* on the Forest

A piece of forest land must be cleared to construct a *pan boroj*, and as long as betel leaf cultivation continues at that location, the forest cover will not return. This is another way that *pan boroj* affect the forest.

Table 3.2 summarizes the construction of all 152 *pan boroj* in Marishbunia over time. Note that these data include *boroj* that are not currently in operation. Additionally, a certain number of *boroj* are expected to be missing from this listing, because a site used by a *boroj* may not be obvious after a few years of disuse, and because evidence of a *boroj* constructed in a flat area between rice fields may be easily obliterated.

Only a few *pan boroj* existed prior to the 1960s (Table 3.2). These long-lasting *boroj* are all located along the beach. *Boroj* that were constructed in the 1970s are few in number, but some of them were built in the forested area. However, all of the *boroj*, except one, that were constructed in the 1980s were built in the forest.

Table 3.3 Location, size, and number of *pan boroj*

Location	Size (ha)	Number of *boroj*
Plain	2.2	41
Slope	7.0	111
Total	9.2	152

The *pan boroj* built in the 1990s show a similar trend to those built in the 1980s. They were mainly built in the forest to fill gaps left by earlier constructions, while a few *boroj* were also built on the flatlands. During the last period (the 2000s and after), almost half of the existing *boroj* (68) were built, and they occupy the entire area of gentle slope up to the edge of the hill, and some of them were even built on the hill.

Table 3.3 shows all of the 152 existing *pan boroj* in the village. Among the 152 *boroj*, 41 are located on the flatlands and 111 are located on the former forested area. These 111 *boroj* occupy at least 7.0 ha of land, on which forest cover will not return. This area is equivalent to 5.4% of the total area (130 ha) in Marishbunia that was once forested.

3.4 Conclusions

This chapter analyzed three issues regarding betel leaf cultivation: (1) the status of betel leaf cultivation in the spectrum of subsistence activities in Marishbunia; (2) the impacts of materials used to construct *pan boroj* in forests; and (3) the area of forests denuded by the placement of *pan boroj*. Regarding subsistence activities, because of the limitation of the agricultural land, rice production in Marishbunia is too low to be the sole source of income for the villagers. Therefore, they must supplement their income from rice production to make ends meet. Thus, betel leaf cultivation may be a means of supplementing their income, because the potential income per unit of land used for betel leaf cultivation may be much higher than that used for rice cultivation.

The use of forest materials for betel leaf cultivation accounts for 4% of the annual estimated biomass production of the forest of Marishbunia village. Moreover, 111 *pan boroj* (7 ha in area) were constructed in formerly forested areas. This means that these *pan boroj* occupy approximately 5% of the forest area of the village. These two types of impacts by betel leaf cultivation together affect approximately 10% of the forest area. The parameters used in the above estimates are by no means fine-tuned; however, these estimates indicate that betel leaf cultivation substantially influences the state of the forest in the village.

References

Bae M, Watanabe C, Inaoka T, Sekiyama M, Sudo N, Bokul MH, Ohtsuka R (2002) Arsenic in cooked rice in Bangladesh. Lancet 360:1839–1840

Choudhury UK, Rahman MM, Mandal BK, Paul K, Lodh D, Biswas BK, Basu GK, Chanda CR, Saha KC, Mukherjee SC, Roy S, Das R, Kaies I, Barua AK, Palit SK, Quamruzzaman Q, Chakraborti D (2001) Groundwater arsenic-contamination and human sufferings in West Bengal, India and Bangladesh. Environ Sci 8(5):393–145

Flora MS, Mascie-Taylor CGN, Rahman M (2012) Betel quid chewing and its risk factors in Bangladesh adults. WHO South-East Asia J Public Health 1(2):169–181

Ghosh S, Maiti A (2011) Betel leaf cultivation – a potential crop for sustainable income. An in-depth study on economy of betel leaf cultivation and marketing with reference to Nadia District, West Bengal, India, 120 p

Gupta PC, Warnakulasuriya S (2002) Global epidemiology of areca nut usage. Addict Biol 7:77–83

Kumar N (1999) Betel vine (Pipper betel L.) cultivation: a unique case of plant establishment under anthropogenically regulated microclimatic conditions. Indian J Hist Sci 34(1):19–32

Rooney DF (1993) Betel chewing traditions in South-East Asia. Oxford University Press, Kuala Lumpur

Smith NM, Lee R, Heitkemper DT, Cafferky KD, Haque A, Henderson AK (2006) Inorganic arsenic in cooked rice and vegetables from Bangladeshi households. Sci Total Environ 370:294–301

Tani M, Jahiruddin M, Egashira K, Kurosawa K, Moslehuddin AZM, Rahman MZ (2012) Dietary intake of arsenic by households in Marua village in Jessore. J Environ Sci Nat Resour 5(1):283–288

Tsuruta H, Tani M, Rahman MZ (2012) The study on the impact of betel leaf cultivation on the local forest in the Teknaf Peninsula. In: Proceeding of 3rd international conference on environmental aspects of Bangladesh, pp 51–53

Varier PS (1995) Indian medicinal plants a compendium of 500 species, vol 4. Orient Longman Publication, Chennai, pp 279–280

Watanabe C, Kawata A, Sudo N, Sekiyama M, Inaoka T, Bae M, Ohtsuka R (2004) Water intake in an Asian population living in arsenic-contaminated area. Toxicol Appl Pharmacol 198:272–282

Chapter 4
Socioeconomic Status of Betel Leaf Farmers

H. Tsuruta

Abstract This chapter sheds light on the socioeconomic status of betel leaf farmers within the wider society. Because betel leaf cultivation is one of the factors leading to deforestation, a study of the socioeconomic status of betel leaf farmers is important for understanding the characteristics of actors who contribute to deforestation through their dependence on forests for supporting their livelihoods in the study area. To ascertain the socioeconomic status of betel leaf farmers, we first present an overall socioeconomic profile of this society. Next, we situate betel leaf farmers within this society. Consequently, we discuss the socioeconomic issues that relate to betel leaf farmers within the society, shedding light on the socioeconomic status of this group. Betel leaf cultivation is prevalent in the western part of the Teknaf peninsula, especially in the Baharchara union. Therefore, we selected three villages in the Baharchara union, namely, Marishbunia, Jahajpura, and South Shilkhali, as our study area. Data were collected from 580 households, with households divided into seven categories based on their main income source. In this society, which evidences economic disparity associated with limited farm land, low education levels, an increasing population, and the expansion of illegal settlements, betel leaf farmers occupy a low class position that accounts for a major proportion of the society. Thus, the socioeconomic status of betel leaf farmers, who constitute the majority of new illegal forest dwellers, is characterized by poverty and low education levels.

Keywords Betel leaf cultivation • Socioeconomic status • Household survey • Teknaf peninsula • Major income source

4.1 Introduction

This chapter discusses the socioeconomic status of betel leaf farmers. Because betel leaf cultivation is one of the factors that lead to deforestation (Tani et al. 2012), studying the socioeconomic status of betel leaf farmers is important, as many people depend on the forest for their livelihoods and subsistence in the study area. To clarify

H. Tsuruta (✉)
Kokusai Kogyo Co. Ltd., Tokyo, Japan
e-mail: jacobie.landscape@gmail.com

M. Tani, M.A. Rahman (eds.), *Deforestation in the Teknaf Peninsula of Bangladesh*, DOI 10.1007/978-981-10-5475-4_4

the socioeconomic status of betel leaf farmers, first, the socioeconomic conditions of the society were examined. Second, betel leaf farmers were categorized in relation to socioeconomic issues, and their socioeconomic status was evaluated.

4.2 Methodology

Betel leaf cultivation is mainly concentrated on the western side of the Teknaf peninsula, especially in the Baharchara union (Tsuruta et al. 2012). Therefore, three villages in the Baharchara union, Marishbunia, Jahajpura, and South Shilkhali, were selected for the study. A series of household surveys covering all of the villages was conducted from 2010 to 2012. To supplement these surveys, another survey was conducted in September 2013. A total of 580 households in the three villages were surveyed, including 159 households (76% of the households) in Marishbunia, 164 households (55% of the households) in Jahajpura, and 257 households (86% of the households) in South Shilkhali (Table 4.1). Factors related to household socioeconomic status include the following: household income, the education level of the head of household, settlement duration, engagement rate in farming activity, location of the house relative to the Teknaf Wildlife Sanctuary (TWS) (inside/outside), and the legal status of the residents inside the TWS.

To clarify the socioeconomic status of betel leaf farmers, all of the households were divided into seven groups according to their main livelihoods: farming (except betel leaf cultivation), betel leaf cultivation, fishing, day labor, working abroad, business, and others. It was common for one household to be engaged in more than one income-generating activity. In such cases, households were classified according to the livelihood that contributed the highest amount to the household income.

Data included quantitative variables (household income, education level of the head of household, and settlement duration) and qualitative variables (engagement in farming and legal status of the house lot). A Kolmogorov–Smirnov test was used to confirm the normality of distribution. To compare the qualitative variables, the Mann–Whitney U test was used to compare the means among the seven livelihood groups.

Table 4.1 Household survey in the study area

Village	Period	Survey contents	Sampled households	Total households
Marishbunia	Sep 2010	Household income	159	210
	Sep 2013	Education level of household		
Jahajpura	Sep 2011	Settlement duration	164	300
	Sep 2013	Engagement rate in farming		
South Shilkhali	Sep 2012	Location of house lots	257	300
	Sep 2013	Legal status of house lots inside the TWS		

4.3 Results and Discussion

4.3.1 General Characteristics of the Households

Location and legal status of settlement: Although settlements inside the TWS are prohibited, except for people registered as forest villagers, there were many houses inside the TWS. A total of 381 houses (66%) were counted inside the TWS, while the remaining 199 houses (34%) were located outside the TWS (Table 4.2). Among the houses located inside the TWS, 200 houses (52%) were categorized as legal, while the other 181 houses (48%) were illegal (Table 4.2).

Settlement duration: It was not possible to determine the length of settlement for 30 households; therefore, these households were excluded from the length of settlement analysis. In general, the number of households gradually decreased as the length of settlement increased, except for the oldest category (more than 100 years) (Table 4.3). The largest category consisted of 124 (23%) households that were established approximately 100 years ago, while the second largest category consisted of 102 (19%) newly established (0–9 years) households (Table 4.3). The increasing trend for new houses to be built in the TWS in recent years may be due to increases in population and migration from other locations.

Table 4.2 Distribution of the respondents according to location and residence status of the houses in the study area

Location	Respondent no. (%)	Status	Respondent no. (%)
Inside the TWS	381 (66)	Legal	200 (52)
		Illegal	181 (48)
Outside the TWS	199 (34)	Legal	199 (100)
Total	580 (100)		

Table 4.3 Distribution of the respondents according to their settlement duration

Settlement years	No. of respondents	% of respondents
0–9	102	19
10–19	90	16
20–29	70	13
30–39	49	9
40–49	44	8
50–59	27	5
60–69	29	5
70–79	7	1
80–89	5	1
90–99	3	0
100+	124	23
Total	550	100

Table 4.4 Distribution of the respondents according to education level of household head

Education level (schooling years)	No. of respondents	% of respondents
No education (0)	366	65
Primary (1–5)	135	24
Secondary (6–10)	45	8
Above secondary (>10)	14	3
Total	560	100

Table 4.5 Distribution of the respondents (households) according to household annual income

Income group	No. of respondents	% of respondents
Very low (<50,000 BDT)	151	26
Low (50,001–100,000 BDT)	202	35
Medium (100,001–150,000 BDT)	80	14
High (150,001–200,000 BDT)	43	7
Very high (>200,001 BDT)	100	18
Total	580	100

Education level of heads of household: Educational attainment in the study area is much lower than the national average (BBS 2009), as school attendance is very low and there are few educational institutions. Because the education level of 20 households was unknown, these households were excluded from the education level analysis. The households were divided into four groups: no education, primary (1–5 years of education), secondary (6–10 years of education), and above secondary education (>10 years of education). Among the 560 heads of household, 366 (65%) had no formal education. One hundred and thirty-five heads of household (24%) had primary education, whereas a few had secondary (8%) and above secondary levels (3%) of education (Table 4.4).

Household income: The average annual income of the household in the study area was 134,700 Bangladeshi Taka (BDT), which was lower than the average (150,000 BDT) of all rural areas in Bangladesh (BBS 2010). Based on the annual income, the households were categorized into five groups: very low (<50,000 BDT), low (50,001–100,000 BDT), medium (100,001–150,000 BDT), high (150,001–200,000 BDT), and very high (>200,001 BDT). Among the categories, most of the households (35%) were in the low-income group, followed by the very low- (26%) and very high (18%)-income groups (Table 4.5). Thus, a substantial number of households in the study area were poor.

Farmland ownership: Among the households, 357 households (62%) owned farmland, while 223 (38%) did not. Although all of the households have homesteads, but these were not considered to be farmland (Table 4.6).

Table 4.6 Distribution of the respondents according to farmland ownership

Farmland ownership	No. of respondents	% of respondents
Own farmland	357	62
No farmland	223	38
Total	580	100

Table 4.7 Distribution of the respondents according to major income-generating activities

Livelihood activity group	No. of respondents	% of respondents
Farming (except betel leaf cultivation)	38	7
Betel leaf cultivation	145	25
Fishing	40	7
Day labor	199	34
Working abroad	58	10
Business	71	12
Other	29	5
Total	580	100

4.3.2 *Characterization of Betel Leaf Farmers in Relation to Other Livelihood Activities*

Distribution of the households based on their livelihood activities: Day labor was the largest livelihood activity category (199 households, 34%), followed by betel leaf cultivation (145 households, 25%), business (71 households, 12%), working abroad (58 households, 10%), fishing (40 households, 7%), non-betel leaf farming (38 households, 7%), and others (29 households, 5%) (Table 4.7). Most day laborers engaged either in farming (betel leaf cultivation and others) or in fishing.

Household income: The annual incomes of different livelihood groups are shown in Table 4.8. The annual income of the day labor group was significantly lower than those of the other livelihood groups. This means that day laborers are the poorest group in society. The betel leaf cultivation group was the second poorest group, although there was a significant difference in mean income between the day labor and betel leaf cultivation groups. The fishing and working abroad groups were the wealthiest groups in society (Table 4.8). There were significant differences in mean income between the day labor and working abroad groups, the day labor and business groups, the day labor and betel leaf cultivation groups, the farming and fishing groups, the farming and working abroad groups, the betel leaf cultivation and fishing groups, and the betel leaf cultivation and working abroad groups.

Education level of heads of household: The education levels of the livelihood activity groups were generally poor. Because the rate of school attendance was very low in the study area, regardless of the livelihood activity groups, there were few significant differences in the means. Among the livelihood activity groups, people engaged in business were more educated (3.8 years of schooling), presumably because this occupation requires mathematics and other accounting skills. The

Table 4.8 Annual income (BDT) of the livelihood activity groups

Livelihood activity group	Mean	SD	Minimum	Maximum
Farming	122,300	115,900	7800	501,600
Betel leaf cultivation	118,200	105,400	21,600	684,000
Fishing	234,600	191,000	36,000	979,200
Day labor	73,500	59,300	7200	495,000
Working abroad	277,900	198,700	24,000	960,000
Business	152,700	153,700	6000	720,000
Other	141,700	149,900	12,000	710,600
Average	134,700	140,700	6000	979,200

SD standard deviation

Table 4.9 Education level (years of schooling) of heads of household of different livelihood activity groups

Livelihood activity group	Education level	SD	Minimum	Maximum
Farming	2.3	3.7	0	12
Betel leaf cultivation	1.7	3.0	0	12
Fishing	2.3	3.4	0	16
Day labor	1.2	2.0	0	8
Working abroad	2.2	3.1	0	12
Business	3.8	4.4	0	16
Other	2.9	4.5	0	16
Average	2.0	3.2	0	16

SD standard deviation

education levels of the day labor (1.2 years) and betel leaf cultivation groups (1.7 years) were lower, although these differences were not significant. The education levels of the working abroad, fishing, and farming groups were almost the same (Table 4.9). There were significant differences in the education level between the day labor and business groups and the betel leaf cultivation and business groups.

Settlement duration: The settlement duration of the livelihood groups varied remarkably in the study area. On average, people that belong to fishing communities settled relatively earlier (64 years) than the other livelihood activity groups. The farming and betel leaf cultivation groups settled approximately 57 and 47 years ago, respectively.

Initially, fishing and betel leaf cultivation began on a small-scale level, and they were managed by the practitioners themselves. However, these occupations subsequently expanded and required many day laborers. Since then (41 years ago), many people began working as day laborers in the fishing and betel leaf cultivation industries. Later, many people started to work abroad (as of 38 years ago) and in business (as of 33 years ago) (Table 4.10). Among the different livelihood activity groups, the settlement durations of the fishing, farming, and betel leaf cultivation groups were significantly higher than those of the day labor, working abroad, and business groups. There was a significant difference in the settlement duration between the betel leaf cultivation and fishing groups, the betel leaf cultivation and business

Table 4.10 Settlement duration (years) of households of different livelihood activity groups

Livelihood activity group	Mean	SD	Minimum	Maximum
Farming	57	52	0	200
Betel leaf cultivation	47	36	1	200
Fishing	64	36	4	100
Day labor	41	38	0	150
Working abroad	38	33	0	110
Business	33	30	0	110
Other	41	48	0	150
Average	44	38	0	200

SD standard deviation

Table 4.11 Distribution of households according to the location of houses of different livelihood activity groups

	Inside the TWS (respondents)		Outside the TWS (respondents)	
Livelihood activity group	Number	%	Number	%
Farming	25	74	9	26
Betel leaf cultivation	109	82	24	18
Fishing	11	28	29	72
Day labor	87	50	87	50
Working abroad	40	70	17	30
Business	47	68	22	32
Other	22	79	6	21
Total	341		194	
Average		64		36

groups, the fishing and day labor groups, the fishing and working abroad groups, and the fishing and business groups.

Location and legal status of houses: House location is related to the household income source, as workplace location is one of the factors that influence the choice of a house location. People associated with farming (74%) and betel leaf cultivation (82%) tended to live inside the TWS, while fishermen (72%) tended to live mostly outside the TWS, followed by day laborers (50%) (Table 4.11). Among the households living inside the TWS, approximately half (49%) were illegal residents. The proportion of illegal settlement inside the TWS was higher in the farming group (60%), followed by the working abroad (57%), business (53%), day labor (45%), and betel leaf cultivation (44%) groups (Table 4.12). The proportion of illegal settlement in the TWS was high (64%).

Farmland ownership: All of the people (100%) engaged in farming or betel leaf cultivation owned farmland, followed by those in the working abroad (65%), fishing (60%), and business (54%) groups. People other than farmers might own farmland to stay financially solvent. Day laborers had the lowest (34%) percentage of land ownership. On average, 63% of the respondents owned farmland, while the rest did not (Table 4.12).

Table 4.12 Farmland ownership by households of different livelihood activity groups

	Own farmland (respondents)		Do not own farmland (respondents)	
Livelihood activity group	Number	%	Number	%
Farming	34	100	0	0
Betel leaf cultivation	133	100	0	0
Fishing	24	60	16	40
Day labor	59	34	115	66
Working abroad	37	65	20	35
Business	37	54	32	46
Other	9	32	19	68
Total	333		202	
Average	63		37	

4.3.3 *Socioeconomic Status of Betel Leaf Cultivators in Relation to Other Livelihood Groups*

In general, the socioeconomic condition of the betel leaf cultivation group, in terms of annual income, education level, and settlement year, was not satisfactory. Annual income in the betel leaf cultivation group was significantly lower than those of the fishing, working abroad, and business groups, although it was significantly higher than that of the day labor group. Education level was significantly different for the business group. However, the overall education level in the society was poor. Settlement year was significantly different for the fishing and working abroad groups (Table 4.13).

4.3.4 *Socioeconomic Factors That Pushed Local People to Engage in Betel Leaf Cultivation*

Disparity between rich and poor: There was a significant disparity between the rich and poor in the study area. An income source analysis showed that the fishing and working abroad groups tended to earn higher incomes and were more likely to own farmland, while the poorest livelihood group was day laborers, who were less likely to own farmland. The betel leaf cultivation and farming groups also tended to be poor, but their annual incomes were higher than that of the day laborer group.

Lack of education: The education level in this area was quite low compared with the national average of rural areas. This might be because the number of schools is small and because the demand for day labor is high, as fishing and farming are labor intensive. Hence, many young males are engaged in activities as day laborers and do not attend school. Moreover, schools are far from each other. The income source analysis showed that the education level of heads of household was lower in the low-income groups (day labor, farming, and betel leaf cultivation) than in the high-

Table 4.13 Relationship between betel leaf cultivators and other livelihood groups in terms of annual income, education level, and settlement year in the study area

Livelihood activity group	Annual income	Education level	Settlement year
Farming	1.000	1.000	1.000
Fishing	0.000***	1.000	0.048*
Day labor	0.000***	1.000	0.290
Working abroad	0.000***	1.000	0.521
Business	0.000***	0.002**	0.048*

Level of significance: * 0.05, ** 0.01, and *** 0.001 according to Mann–Whitney U tests

income groups. Business people had the highest education level, which may be because conducting business requires mathematics skills and other knowledge.

Population increase and illegal settlement: Population increase might be a root cause of deforestation, because it requires more forest resources for the construction of new houses, fuelwood for cooking, and materials for betel leaf cultivation. Furthermore, more land is required for the construction of new houses, and new houses were built mostly in the TWS.

4.4 Conclusions

The economic status of betel leaf farmers was characterized as poor, as they were less educated and more likely to dwell illegally in the forest. This study focused on the major livelihood activity groups of society. Thus, an in-depth study of betel leaf farmers should be conducted to thoroughly analyze their socioeconomic status, profitability, productivity, the marketing systems they employ, and their impacts on forest and other natural resources.

References

BBS (2009) Updating poverty maps of Bangladesh, Bangladesh bureau of statistics. People's Republic of Bangladesh, Dhaka

BBS (2010) Report of the household income & expenditure survey 2010, Bangladesh Bureau of Statistics. Dhaka, People's Republic of Bangladesh

Tani M, Rahman MZ, Asahiro K (2012) The study of the influence of betel leaf cultivation on forest loss in the Teknaf Peninsula, Bangladesh, Geijutu Kougaku. J Des 16:1–10

Tsuruta H, Tani M, Rahman MZ (2012) The study on the impact of betel leaf cultivation on the local forest in the Teknaf Peninsula. In: Proceeding of 3rd international conference on environmental aspects of Bangladesh, pp 51–53

Chapter 5
Economic Impacts of Betel Leaf Cultivation on Farmers

M. Sakamoto

Abstract Substantial deforestation has occurred in the Teknaf peninsula. This chapter presents a detailed case study on the effects of betel leaf cultivation, as one of the likely causes of deforestation, on local people's livelihoods in the village of Marishbania in the Teknaf peninsula. Grown as a cash crop, betel leaf is a popular masticant across South Asia. The practice of betel leaf cultivation in the TWS entails the construction of a *pan boroj*, which is a shed constructed out of forest resources that offers shade for protecting the leaves from sunlight. This shed requires annual renovation. In September 2010, a questionnaire was administered among all 207 households in the village. The compiled data revealed differences in household characteristics, which were analyzed in detail, with groups being categorized according to when villagers first began cultivating betel. The results of a statistical analysis conducted on households' income levels revealed statistically significant differences in income levels between households that began cultivating betel during the last 10 years and those that began cultivating it earlier. A particularly notable finding was that betel leaf cultivation was not necessarily associated with increased incomes for those who have been cultivating it during the last 10 years.

Keywords Betel leaf • Cash crop farming • Crop selection • Questionnaire survey • Income disparity

5.1 Introduction

Substantial deforestation is occurring on the Teknaf peninsula, and sufficient measures to mitigate deforestation have not yet been implemented. The possible causes of deforestation are firewood collection by locals, population migration into forested areas, and betel leaf cultivation. Betel leaf, a cash crop broadly popular among people in South Asia, is a popular mastication, usually chewed with a slice of betel nut and lime. The leaf is used in social, cultural, and religious events for hospitality, and

M. Sakamoto (✉)
Graduate School of Frontier Sciences, The University of Tokyo, Chiba, Japan
e-mail: m-sakamoto@k.u-tokyo.ac.jp

M. Tani, M.A. Rahman (eds.), *Deforestation in the Teknaf Peninsula of Bangladesh*, DOI 10.1007/978-981-10-5475-4_5

it also has medicinal value (Guha 2006). In Bangladesh, it is widely cultivated in Sylhet, Moulvibazar, Jessore, Khulna, Kustia, Bagerhat, Satkhira, Narail, Bhola, Barisal, Faridpur, Rajshahi, Rangpur, Gaibandha, Pabna, Cox's Bazar, and the greater Chittagong district (BBS 2010; Fila et al. 2006). In some regions (particularly the northern part), local people use a traditional method to grow betel leaf on trees, which is different from that used for betel leaf cultivation on land. Tree-based betel leaf cultivation is suggested to be a productive and sustainable agroforestry system (Haider et al. 2013). However, Tani et al. (2013) reported that the use of forest trees to construct *pan boroj*, which are used to grow betel leaf, negatively impacts the forest environment and contributes to deforestation on the Teknaf peninsula. This chapter uses a village (Marishbania) on the Teknaf peninsula as a case study, and it closely examines the effects of betel leaf cultivation on the livelihoods of the local people in the village. Using questionnaire data to survey the beliefs and individual attributes of locals, differences in household characteristics are analyzed in detail, with groups classified according to when villagers started cultivation (Sakamoto et al. 2014).

5.2 Methodology

The questionnaire survey was conducted in September 2010 and the pan cultivation survey was conducted in March 2011. Two hundred and seven households in the village participated in face-to-face interviews using a standardized interview schedule. The questions required "yes"/"no" answers as well as numerical information and names. The attributes of the households in terms of occupation, annual income, and year of involvement in betel leaf cultivation are shown in Figs. 5.1, 5.2, and 5.3.

In 2010, almost half of the households were farmers (Fig. 5.1), and the annual income of most of the households (58%) was 50,000 Bangladeshi Taka (BDT) or less per year (Fig. 5.2). The mean household annual income is lower than that of non-

Fig. 5.1 Occupation of the head of household

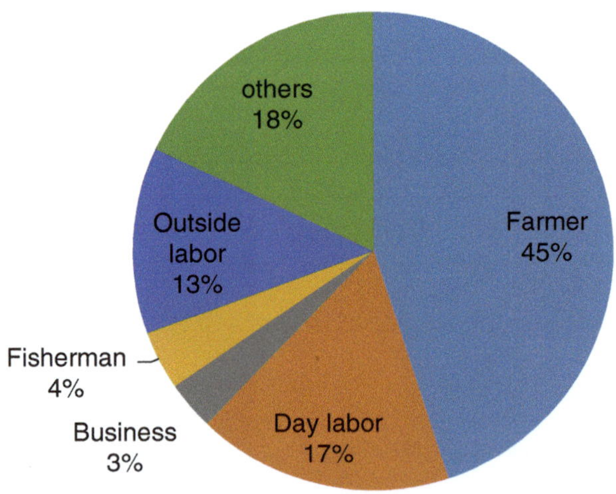

Fig. 5.2 Annual income (BDT) of the household

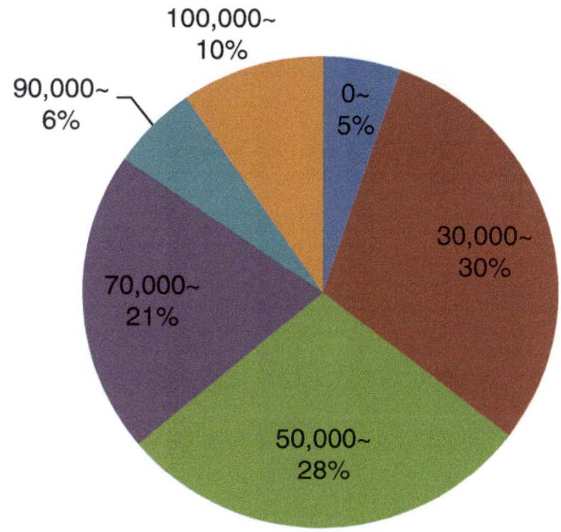

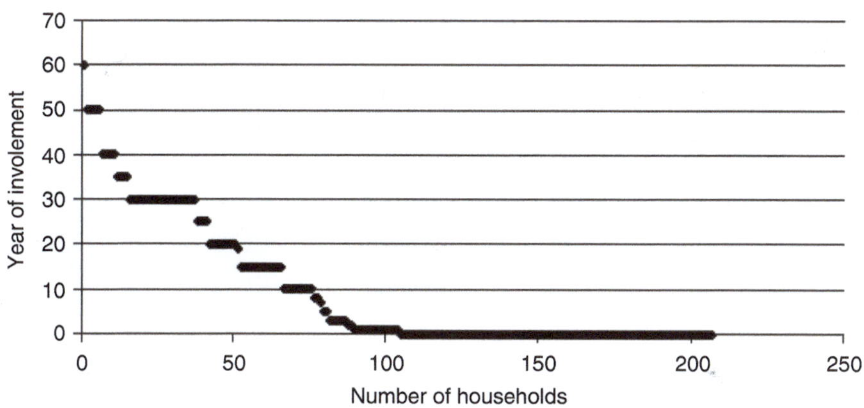

Fig. 5.3 Years of involvement in betel leaf cultivation

manufacturing workers in urban areas of the country. In Fig. 5.3, the 207 households are plotted corresponding to their years of engagement in betel leaf cultivation. Households that did not cultivate betel leaf were plotted at zero. At the time of the survey, about half (104) of the households cultivated betel leaf, while 103 did not. Figure 5.3 shows that betel leaf cultivation has rapidly expanded in the last 30 years.

The responses of the households regarding their environmental attitudes are summarized in Figs. 5.4 and 5.5. Figure 5.4 shows answers to the question, "Compared to 10 years ago, what kind of change do you notice in the natural/physical environment?" Figure 5.5 shows answers to the question, "What is the cause of that environmental change?" Approximately 94% (194 out of 207 households) responded that they noticed a decrease in the number of trees, suggesting that degradation has been underway during the past 10 years. The vast majority of respondents chose

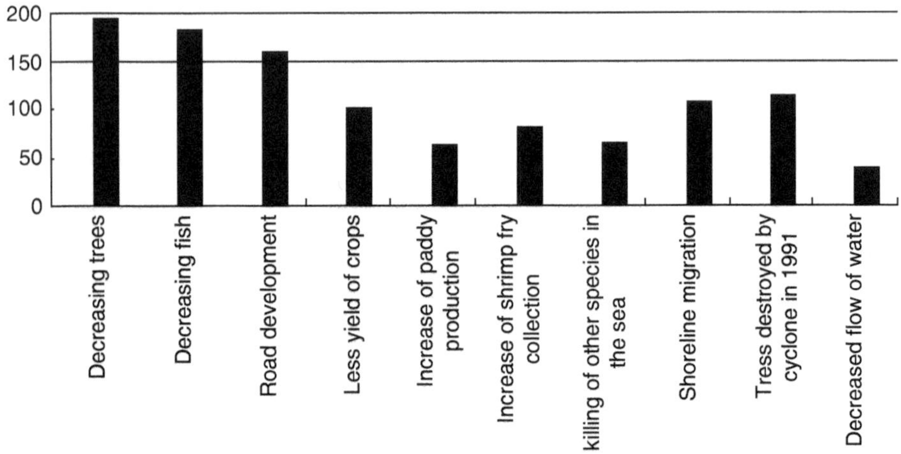

Fig. 5.4 Responses of the households regarding the kinds of changes they noticed in the natural/physical environment

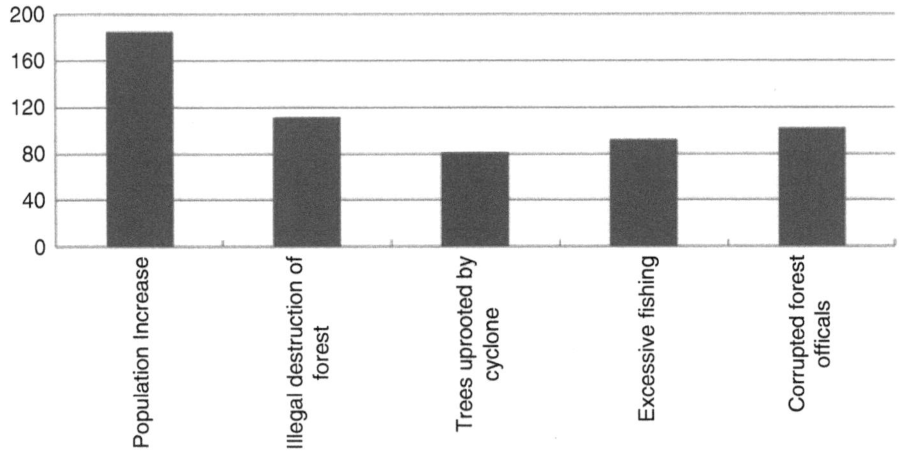

Fig. 5.5. Responses of the households regarding the causes of environmental change

population increase as a major cause of environmental change, and approximately half pointed to illegal logging as a cause of environmental change (Fig. 5.5). Thus, aside from natural factors, the respondents reported that anthropogenic activities caused deforestation. These results contradict the current thinking that the increasing numbers of *pan boroj* are responsible for the continuing deforestation.

The economic and social factors associated with betel leaf cultivation in the village were statistically analyzed using data collected in the survey that was conducted in September 2010. The households were divided into two groups: (1) those who cultivated betel leaf and (2) those who did not. A series of *t*-tests evaluated

Table 5.1 Household characteristics

Classification	Variable
Personal property	Number of family members
	Annual salary
	Annual salary/number of family members
	Amount of loan
	Number of cows
	Number of goats
	Number of chickens
Attitude to environment	Attitude to improving environment
	Complaint about actual situation
	Recognition of environmental change

Table 5.2 Questions used to score the variables measuring attitude regarding the environment

Variable	Questions
Attitude to improving environment	Access to forest should be regulated
	Trees should be planted
	Local people should act for reforestation
	Illegal destruction of forest is one cause of environment change
Complaint about actual situation	Less grass, more cost of housing materials
	Need to buy wood from Teknaf
	Feel unwell due to environment degradation
	Insufficient food
Recognition of environment change	Fish decreasing
	Road developed
	Wild animals killed
	Less yield of crops
	Temperature of climate increased

whether the means of the two groups differed with respect to the ten household characteristics. Table 5.1 lists the tested variables.

The three "attitude regarding the environment" variables are composite indices of answers to several questions. Each respondent's score on an index is the sum of his or her "yes" responses to the questions that comprise that index. For example, scores on "attitude regarding improving the environment" (Table 5.2, panel (1) ranged from 0 (no "yes" responses to the listed questions) to 4 (all "yes" responses to the listed questions). Under this scoring system, the higher the score, the greater the willingness to improve the environment, the more complaints about the environmental situation, and the higher the perception of the severity of environmental change. Table 5.2 lists the questions used to score the three index variables classified as attitude regarding the environment.

5.3 Results and Discussion

5.3.1 Relationship Between Personal Attributes and the Choice of Betel Leaf Cultivation at the Present Time

The differences in "personal property" characteristics and in "attitude regarding the environment" (Tables 5.1 and 5.2) between betel leaf cultivators and non-cultivators were assessed by a *t*-test. Table 5.3 shows the statistically significant differences in mean values between the two groups. For simplicity in the following tables, "with" indicates the betel leaf cultivators and "without" indicates the non-cultivators.

Of the ten *t*-tests, Tablet 5.3 shows that two variables exhibited statistically significant differences between the two groups. First, the mean family size of the cultivators was significantly higher than that of the non-cultivators (7.33 versus 5.17, respectively) at a significance level of $P < 0.05$. Betel leaf cultivation requires intensive labor, and it may be that large families are better able than small families to take on such work. Conversely, a family that cultivates betel leaf and has a need for field workers may be more inclined than a non-cultivating family to increase its size.

Complaints about the actual environmental situation also differed significantly, on average, between the two groups at the level of $P < 0.1$, with betel leaf cultivators averaging relatively more complaints. Betel leaf cultivators have recently been obliged to obtain trees and thatch from the market instead of from the forest at no cost. Therefore, they actually have to pay more to maintain their living environment. This may have led to more complaints from cultivators about the actual environmental situation.

5.3.2 Relationship Between Personal Attributes and the Choice of Betel Leaf Cultivation in the Past

Based on the observation of the trend in the years since households started betel leaf cultivation (Fig. 5.3), we classified the years, into four time-period groups, and performed a series of *t*-tests to assess the differences in household characteristics in 2010 (Table 5.1) between the two groups (cultivators who started in each of the four periods and the rest of the households). The analysis showed a relationship between

Table 5.3 Differences between betel leaf cultivators ("with") and non-cultivators ("without")

Variable	Significance	Mean	
Number of family members	0.027	With	7.3269
		Without	5.1650
Complaint about actual situation	0.071	With	2.0385
		Without	1.7476
			Sample size: with,104; without, 103

Table 5.4 Differences between cultivators and non-cultivators according to the period in which they started betel leaf cultivation

Period	Variable	Significance	Mean		Sample size
First	NA	NA	NA		First, 15; others, 192
Second	NA	NA	NA		Second, 26; others, 181
Third	Number of goats	0.034	Third	0.3200	Third, 25; others, 182
			Others	0.6978	
Fourth	Number of goats	0.066	Fourth	1.1842	Fourth, 38; others, 169
			Others	0.5325	
	Annual salary	0.075	Fourth	128,735	
			Others	354,683	
	Annual salary/ number of family	0.059	Fourth	19,739	
			Others	49,075	
					Sample size: with, 104 without, 103

the timing of the first cultivation and differences in the household characteristics in 2010, which helps our understanding of the characteristics of households that started betel leaf cultivation at different periods.

The years in which involvement began were grouped into periods as follows: (1) 1975 and earlier, (2) 1980–1989, (3) 1990–1999, and (4) 2000–2010. Because the respondents' answers are basically 10-year steps, we have adopted these classification periods by clustering their answers.

Table 5.4 shows the means of the variables that differed significantly between the two groups in each period among the variables listed in Table 5.1. None of the variables differed significantly between the first and second periods. In the 1990–1999 (third) period, the number of goats differed significantly between the two groups (groups ($P < 0.050$), as the mean of the group of households that started betel leaf cultivation in this period was smaller than the rest of the households).

In the fourth period, the number of goats, the annual income, and the annual income/number of family member's ratio differed significantly between the two groups ($P < 0.1$). While those who first cultivated from 1990 to 1999 (the third period) had fewer goats, on average, than the rest of the households, those who first cultivated from 2000 to 2010 had twice as many goats, on average, as the rest of the households. However, the average annual income of the cultivators as well as the annual income/number of family member's ratio of the cultivators were lower than that of the rest of the households. Based on these results, the actual situation of households that started betel leaf cultivation in the fourth period appeared to be significantly different from that of the group that started before the fourth period, as well as from the group that did not cultivate betel leaf.

5.3.3 Hypothesis on the Mechanism of Expanding Income Disparities

The mean annual income of those who started cultivating betel leaf (128,735 BDT) in the fourth period was significantly lower than that of the rest of the households (354,683 BDT) (Table 5.4). Betel leaf is a cash crop, and its cultivation has rapidly increased, as seen in Fig. 5.3, presumably because people believe that undertaking betel leaf cultivation would increase their household income. Why, then, does the group that started cultivation most recently have a lower income?

To answer this question, we examined the relationship between betel leaf cultivation and increases in income. At first, the households who started cultivation in the first through the third periods (1975–1999) were merged. Their income is compared to the incomes of those who had not started cultivation by 2010. A series of *t*-tests did not find any statistically significant differences in the means of the variables listed in Table 5.1 between the two groups. Therefore, we conclude that the income of the cultivators except those who started in the fourth period is not significantly different from that of non-cultivators at 2010. The question is that if betel leaf cultivation contributed to an increase in income of households that started betel leaf cultivation from the first through the third periods or not. Let us set a hypothesis to be tested for answering this question.

Hypothesis: Betel leaf cultivation contributed to an increase in income of households that started betel leaf cultivation from the first through the third periods.

If the hypothesis is supported, the mean income of betel leaf-cultivating households should have increased from those who began cultivation during the first period to those who began cultivation in the third period. If so, the cultivators might have been initially poorer than the current non-cultivators. Given that betel leaf is a cash crop, and those who started cultivation from the first through the third periods could use forest resources to construct a *pan boroj* at no additional cost, betel leaf cultivation must have been related to income increases. Therefore, the hypothesis is more likely to be true.

Then, let us come back to the first question: why is the mean income of households that started betel leaf cultivation in the fourth period significantly lower than all the other households? Is it because cultivation started so recently and the financial benefits of betel leaf cultivation have not had sufficient time to result in higher incomes? To check this aspect, we grouped the households into every possible combination of the two groups based on their answers to the questions about the year in which they began cultivating betel leaf. That is, the mean of the variables was compared between households that started cultivation 1 year ago and the others, between households that started cultivation 5–10 years ago and the others, between households that started cultivation 15–20 years ago and the others, and so on. A series of *t*-tests was used to compare the means of the variables.

We found that annual income and the annual income-to-number ratio of households did not differ significantly for any possible comparison, except the comparison between households that started cultivation 1–10 years ago versus the others as it was shown in Table 5.4. This result demonstrated that the only statistically significant difference in income was between the groups that most recently began cultivation (which had a lower income) and the other groups (which had higher incomes). If less effect of betel leaf cultivation on income generation is a matter of time, another cluster containing the households who started cultivation in recent 5 years, for example, should have significantly lower income than that of all the other households. Thus, we can conclude that the households that started cultivation in recent 10 years have failed to increase their income and this is not because of time effect.

One reason for this may be that forest resources around the village started to decrease about 10 years ago to the point at which the local people were obliged to obtain trees and thatch from the market. Their incomes were negatively impacted by these costs, which did not exist more than 10 years ago. We were able to find another evidence that supports this reasoning as shown in Table 5.5. Some of the other variables from Table 5.1 were found to be statistically significant in the above analysis through the clustering of any possible combination of groups.

As shown in the table, differences in the means of "attitude regarding improving the environment" and "complaints about the actual environmental situation" were found to be statistically significant at $p < 0.05$, as were the difference in the mean of "recognition of environmental change" at $p < 0.01$ between the clustering of 10–15 years and the others. Additionally, differences between the means of "complaints about the actual environmental situation" were statistically significant at $p < 0.05$, as were the differences in the means of the other two variables at $p < 0.1$ between the clustering of 1–15 years and the others. These findings suggest that

Table 5.5 Differences in attitude regarding the environment between certain groups and others

Starting period	Variable	Significance	Mean		Sample size
10–15 years	Attitude regarding improving the environment	0.012	This group	2.9583	This group, 24
			The others	2.3880	The others, 183
	Complaints about the actual environmental situation	0.025	This group	2.3750	
			The others	1.8306	
	Recognition of environmental change	0.006	This group	3.1667	
			The others	2.5355	
1–15 years	Attitude regarding improving the environment	0.100	This group	2.6923	This group, 52
			The others	2.3742	The others, 155
	Complaints about the actual environmental situation	0.032	This group	2.1923	
			The others	1.7935	
	Recognition of environmental change	0.085	This group	2.8654	
			The others	2.5226	

there might be an environmental change that the households who started cultivation after 10–15 years ago particularly were able to recognize. One reasoning behind the result could be that around 10–15 years ago, namely, 1995–2000, forest resources degraded significantly, and households started buying trees and thatch from the market. The households who started cultivation after that time must have been poor as the other cultivators were so initially, and they were not able to obtain sufficient benefits from cultivation because of extra cost of trees and thatch. They might think that the extra cost is burden and, thus, they came to be more sensitive to environment change. However, they might not be aware of the difference in the cost condition from what the other precedent cultivators had owed. That is why they have kept cultivating it and not been able to escape from poverty trap as the precedent cultivators were able to.

5.4 Conclusions

The results of our analysis revealed differences in the characteristics of households that first cultivated betel leaf at different starting periods, which were most pronounced between households that began betel leaf cultivation in the last 10 years and those who began betel leaf cultivation before that. In particular, we found that betel leaf cultivation is not necessarily related to increase income for those who started cultivation in the last 10 years. However, the number of households that began betel leaf cultivation in the last 20 years or so increased rapidly, and the number of cultivators continues to increase. These fourth-period households may think that betel leaf cultivation contributes to income increases, which was true when people could utilize sufficient forest resources for free; these households may be unaware that times have changed and that such conditions no longer exist.

References

BBS (2010) Report of the household income & expenditure survey 2010. Bangladesh Bureau of Statistics, People's Republic of Bangladesh, Dhaka

Fila P, Haider MR, Mohiuddin M, Zashimuddin M, Alam MK. (2006) Khasia betel-leaf cultivation technique (in Bangla), (Bangladesh Forest Research Institute, Chittagong), 12

Guha P (2006) Betel-leaf: the neglected green gold of India. J Hum Ecol 19(2):87–93

Haider MR, Khair A, Rahman MM, Alam MK (2013) Indigenous management practices of betel-leaf (Piper betle L.) cultivation by the Khasia community in Bangladesh. Indian J Tradit Knowl 12:231–239

Sakamoto M, Tani M, Moriyama M (2014) Examining the process of deforestation by cash crop farming in Teknaf and its impact on inhabitants' livelihoods. Int J Environ 4(2):31–38

Tani M, Zulfikar RM, Asahiro K (2013) The study of the influence of betel leaf cultivation on forest loss in the Teknaf peninsula. Bangladesh Kyushu Univ Inst Repository 16:1–10 (in Japanese)

Chapter 6
Remote Sensing Characterization of Changes in Forest Resources and Betel Leaf Cultivation Through Time

M. Sakamoto, M. Tani, and M. Moriyama

Abstract Obtaining accurate data for monitoring the conditions of resources in developing countries is typically challenging. The reason may be insufficient facilities or, in some cases, disorganization of authorities. These difficulties also apply to procuring data on forest resources in Teknaf. There have been some recent initiatives and/or activities to conserve forests in this area. However, recent data are inadequate for determining the underlying mechanism of deforestation in Teknaf. What is required is time series data for the period before the onset of deforestation. Remote sensing offers a powerful tool for conducting a quantitative analysis of a time series and/or seasonal changes in forest resources. In this chapter, we discuss how we used remote sensing data to show changes in Teknaf's forest resources and assess changes in betel leaf cultivation practices among villagers by combining an analysis of remote sensing data with an analysis of the livelihood status of inhabitants of the village of Marishbunia.

Keywords Remote sensing • NDVI • Chronological vegetation analysis • Deforestation • Human disturbance

M. Sakamoto (✉)
Graduate School of Frontier Sciences, The University of Tokyo, Chiba, Japan
e-mail: m-sakamoto@k.u-tokyo.ac.jp

M. Tani
Department of Environmental Design, Faculty of Design, Kyushu University, Fukuoka, Japan
e-mail: tani@design.kyushu-u.ac.jp

M. Moriyama
Division of Electrical Engineering and Computer Science, Graduate School of Engineering, Nagasaki University, Nagasaki, Japan
e-mail: matsu@rsirc.cis.nagasaki-u.ac.jp

M. Tani, M.A. Rahman (eds.), *Deforestation in the Teknaf Peninsula of Bangladesh*, DOI 10.1007/978-981-10-5475-4_6

6.1 Introduction

It is usually difficult to collect good data when monitoring resource conditions in developing countries. This is because of insufficient facilities and a lack of organization of the authorities. This is also true for data regarding the forest resources in Teknaf. While there have been some recent activities to conserve the forests, current data are insufficient to determine the mechanism that led to deforestation in Teknaf. To do so requires time series data prior to the time at which the deforestation occurred. Remote sensing data analysis is a powerful tool for conducting quantitative analyses of time series and/or seasonal changes of forest resources. In this chapter, we use remote sensing data to show a change of forest resources in Teknaf and analyze the changes in betel leaf cultivation practices among villagers by combining the results of the remote sensing analysis with those of the livelihood status of the villagers in Marishbunia, as was discussed in Chap. 5.

6.2 Methodology

The earth observation satellite Landsat 1 was launched in 1972 (Williams et al. 2006). Landsats 5 and 7 are currently in operation, and their successor, Landsat 8, began operation in April 2013. Landsats 1 through 3 observe a location every 18 days, while the temporal resolution of Landsats 4 through 8 was reduced to 16 days. Landsats 1 through 5 were equipped with a four-channel multispectral scanner (MSS) that detects light ranging from the visible spectrum to near-infrared rays, with a spatial resolution of 80 m. Landsats 4 and 5 also were equipped with a thematic mapper (TM) and are able to detect light from the visible spectrum to thermal infrared rays with a resolution of 30 m (120 m for thermal infrared light).

The Normalized Differential Vegetation Index (NDVI) was combined with Landsat satellite imaging of the study site to evaluate the overall health of the forested areas. The NDVI, which was computed from the satellite-derived surface reflectance of the red and near-infrared spectral regions, was used to identify and estimate the area of photosynthetically active leaves from space (Jones and Vaughan 2010), and for over 40 years, researchers have used it to characterize vegetation. Healthy leaves heavily reflect near-infrared light (NIR) while absorbing large amounts of red light. Using this property, the NDVI is calculated as the ratio of the difference between NIR and red light over their sum, resulting in a value between -1 and $+1$, which indicates the health and density of the vegetation in the analyzed area.

We gathered all 23 satellite images with few clouds from the target region. These consisted of 4 images from Landsat 3 taken with the MSS and 19 images taken with the TM or Enhanced TM Plus. Because NDVI values can vary depending on the season, we limited our coverage to a 1-month period in which the most images were available. We chose to analyze nine images from the period of mid-January to early February, ranging from January 14, 1980, to February 5, 2009.

6.3 Results and Discussion

6.3.1 Forest Condition

Figures 6.1 and 6.2 show the NDVI values for January 14, 1980, and February 5, 2009, respectively. Figure 6.3 shows the difference in the NDVI obtained by subtracting the NDVI value of January 14, 1980, from that of February 5, 2009. The green color in Figs. 6.1 and 6.2 indicates a positive NDVI value, and the blue color indicates a negative value; the blue color in Fig. 6.3 indicates that the NDVI value of February 5, 2009, is higher than that of January 14, 1980, while the red color in Fig. 6.3 indicates that the NDVI value of February 5, 2009, is lower than that of January 14, 1980. The sea zone of Fig. 6.3 is colored white by masking the 0-m elevation zone.

Figures 6.1, 6.2, and 6.3 show that the forest zone on the main part of the Teknaf peninsula decreased from 1980 to 2009, which is indicative of forest degradation. The NDVI values near the ocean are higher than those farther inland, possibly because the 1980 image was captured earlier in the season (January 14) than the 2009 image (February 5). The status of the grass may be different simply because of natural growth between January and February, which results in a higher NDVI value. Generally, we conclude that the Teknaf peninsula was suffering from deforestation during this period.

6.3.2 Relationship Between Betel Leaf Cultivation and Deforestation

To understand the relationship between betel leaf cultivation and deforestation, we focused on Marishbunia, for which we have betel leaf cultivation data for all of the households in the village. In Chap. 5, it was observed that betel leaf cultivation has rapidly expanded since 1980.

Next, QuickBird images from March 11, 2008, were used to calculate the NDVI of the forested area around our study site. QuickBird is a commercial, earth-observing satellite operated by DigitalGlobe (Longmont, CO, USA). With an observation width of 16.5 km, its four-channel multispectral sensor (red, green, blue, and NIR (RGB-NIR)) and its panchromatic sensor have spatial resolutions of 2.44 and 0.61 m, respectively. Figure 6.5 shows the location of the village households of the study site (outlined in red) and the forested area that was analyzed (outlined in white).

The NDVI was calculated using the QuickBird images. Based on the calculated NDVI values, meshes with an index of 0.4 or higher were classified as the forest area. This threshold was used to decrease the effects of the atmosphere on the images, as well as to include sparse tree zones as forest. Because the 2.44-m resolution images were processed using pan-sharpen to obtain 0.61-m resolutions by separating the color and brightness from the RGB-NIR multispectral sensor and

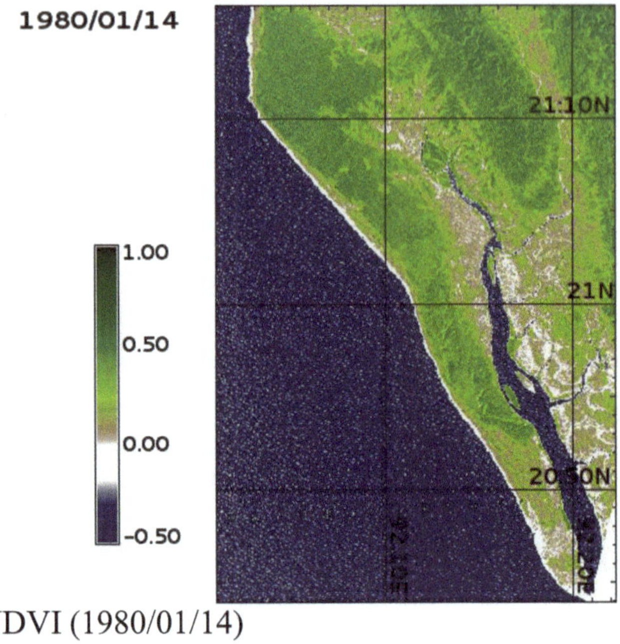

Fig. 6.1 NDVI (January 14, 1980)

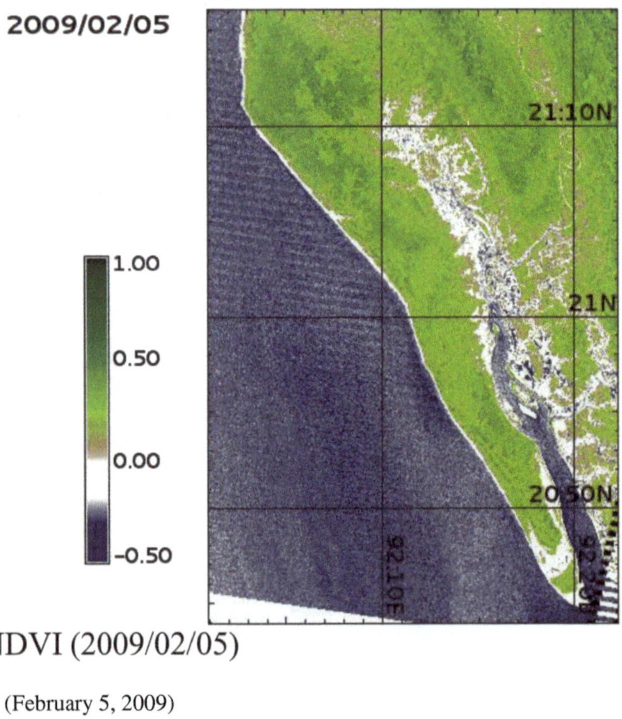

Fig. 6.2 NDVI (February 5, 2009)

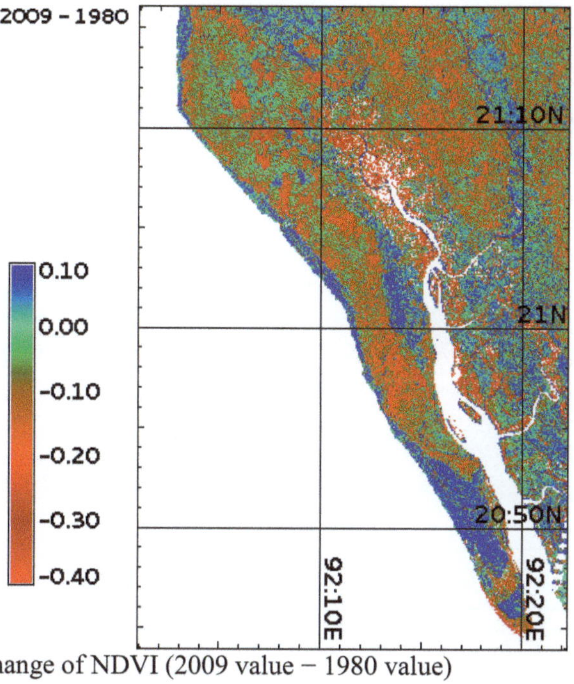

Fig. 6.3 Change of NDVI (2009 value−1980 value)

replacing the brightness with the panchromatic image, they are not as clear as the 0.61-m mesh resolution images.

The resulting NDVI calculation gave a mesh of 10,730 forest points. The shape of the calculated mesh is shown in Fig. 6.5. The black points indicate the mesh that is classified as forest among the meshes located in the white outlined area in Fig. 6.4. Figure 6.6 shows the changes in NDVI values for the nine Landsat images.

According to the slope of the maximum and average values of the NDVI in Fig. 6.6, there was a decline in the health of the surrounding forested areas around the year 2000.

During the 1990s, Landsat made small numbers of observations because of the failure of Landsat 6 and the intermittently extended use of the old Landsat 5. Moreover, this area is often cloudy. Therefore, it was not possible to obtain any fine satellite images from 1989 to 2000. However, even with the lack of data for this period, when observing the trend as a whole, the results are more than sufficient to support our conclusion. The increase of the NDVI after 2000 and 2009 may have resulted from positive impacts of a participatory forestry program (the Nishorgo Support Project) that was undertaken from 1999 to 2000 and from 2008 to 2009 (BFD 2003). There are three social forestry zones around the study site, but they are not located in the forested area that was analyzed (Fig. 6.5). Therefore, the increase of the NDVI after 2000 and 2009 is not a direct effect of the project, but it may have

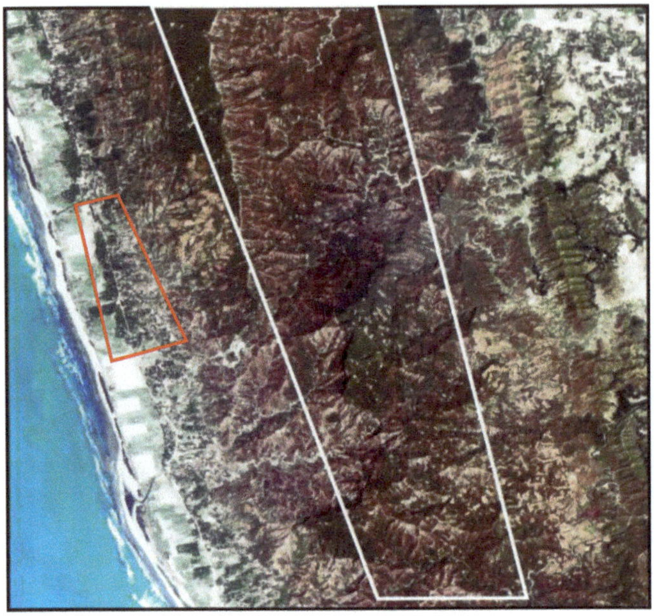

Fig. 6.4 Study site (outlined in red) and forest area (outlined in white)

Fig. 6.5 NDVI extraction location

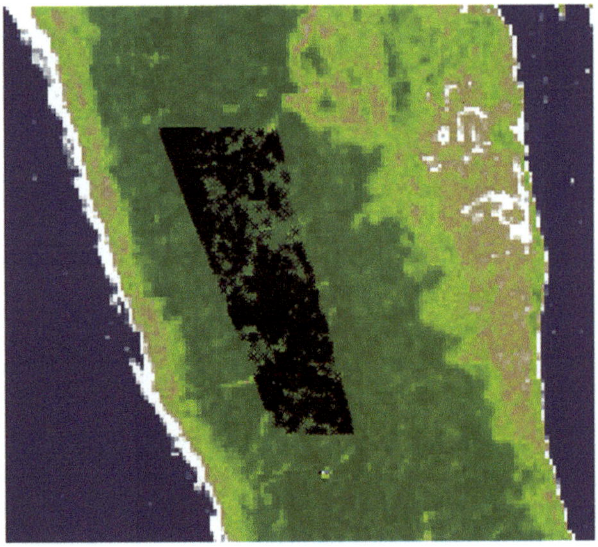

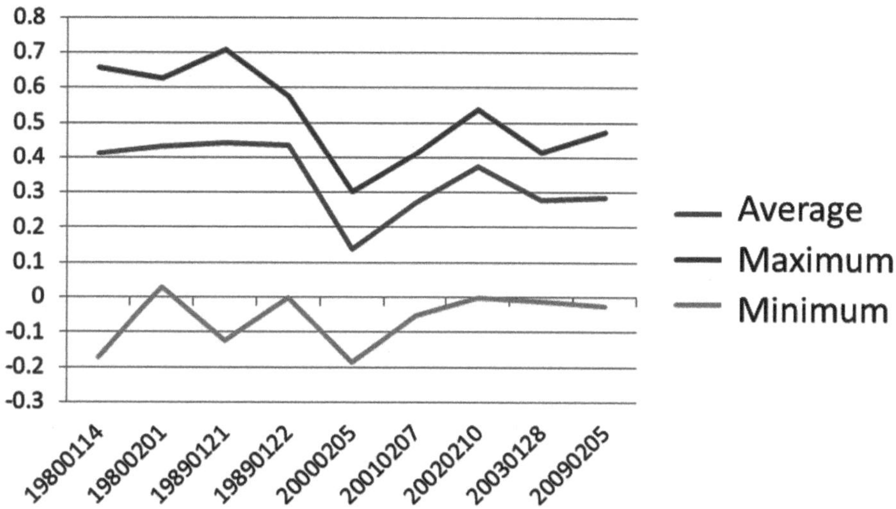

Fig. 6.6 Change in NDVI over time

resulted from an indirect effect of the project, which may have improved people's attitudes regarding forest preservation (Fig. 6.6).

Our results lead us to conclude that the spread of betel leaf cultivation and the depletion of forest resources around Marishbunia are connected and that betel leaf cultivation is a major cause of deforestation. Furthermore, the effects of betel leaf cultivation differ across the time periods in which cultivation began.

6.3.3 Examining the Changes in Betel Leaf Cultivation

Table 6.1 summarizes the results of Chap. 5, which showed that the effects of betel leaf cultivation on household income depend on the time period in which cultivation began, as well as on the time at which the forest resources degraded around the village, which forced the villagers to obtain the resources needed for betel leaf cultivation, such as thatch for the construction of *pan boroj*, from the market. The results showed that because of the timing of the forest degradation, households that started betel leaf cultivation after 2000 have not generated much profit (also see Chap. 5).

The results of the satellite image analysis match the findings in Chap. 5, which are summarized in Table 6.1. Therefore, we can conclude that the period from 1995 to 2000 was a turning point for forest resources, as freely available resources were no longer available. If certain countermeasures had been adopted prior to this time, reforestation may have been conducted more efficiently. The number of households that cultivate betel leaf is increasing, and the use of forest resources that are not related to betel leaf cultivation may become depleted because of a shortage of the forest resources, as buying and selling these resources in the market are promoted.

Table 6.1 Process of betel leaf distribution

First period before 1975	Second period 1980–1985	Third period 1990–1995		Fourth period 2000–2010
Betel leaf cultivation was started experimentally	Betel leaf cultivation was rapidly disseminated among the poor people. It contributed to increased income	Betel leaf cultivation was disseminated among the poor people. It contributed to increased income	Turning point of the forest resource	Betel leaf cultivation was disseminated among the poor people. It did not contribute to income increase

6.4 Conclusions

Because betel leaf cultivation did not produce much financial benefit from 2000 to 2010 compared with the earlier periods, and because it consumes forest resources, new betel leaf cultivation should be restricted. However, the hypothesis test suggested that households that decided to cultivate betel leaf were poor when they began cultivation, and therefore, if betel leaf cultivation were restricted, poor households likely would suffer, because they would be denied a perceived option to earn additional income. Other options may prove to be more beneficial. One possible measure would be to tax the use of forest resources and to reallocate the resulting tax revenues to the local poor through employment opportunities that contribute to reforestation, such as patrolling the forest and planting trees. Environmental education via workshops may increase local appreciation of the forests and encourage villagers to participate in forest conservation activities.

References

BFD (2003) Social forestry: Bangladesh Forest Department. Ministry of Environment and Forest Government of Bangladesh, Bangladesh. Available from http://www.bforest.gov.bd/index.php/forest-management/social-forestry

Jones HG, Vaughan RA (2010) Remote sensing of vegetation: principles, techniques, and applications. Oxford University Press, Oxford

Williams DL, Goward SN, Arvidson TJ (2006) Landsat: yesterday, today, and tomorrow. Photogramm Eng Remote Sens 72(10):1171–1178

Chapter 7
Fuelwood of Teknaf Forests: Marketing, Distribution, and Concomitant People

S.M.A. Ullah and J. Tsuchiya

Abstract On the Teknaf Peninsula, fuelwood extraction is important both as a means of livelihood for the local population and as a cause of deforestation resulting from overharvesting. This chapter focuses on describing the marketing and distribution of fuelwood and characterizing the local people involved in these activities. Fuelwood collection is prevalent in Teknaf, and a significant number of households harvest fuelwood themselves. Poverty, lack of alternative cooking fuel, and easy access to nearby forests are the main reasons for high levels of fuelwood dependence in this area. In addition to depending on fuelwood to meet demands for cooking energy, many households also rely on fuelwood collection, distribution, and marketing as a means of living. Thus, from the perspective of sustainable forest management, it is important to acquire an understanding of fuelwood-related issues that impact on forests. There are four major markets or bazars for fuelwood in Teknaf Upazila: Shamlapur Bazar, Teknaf Bazar, Nhilla Bazar, and Whykong Bazar. Fuelwood prices vary across these four bazars, being lowest in Shamlapur Bazar (6.35 BDT kg^{-1}) and highest in Nhilla Bazar (8.87 BDT kg^{-1}). Of the surveyed fuelwood vendors, 63% were male and 37% were female. Their ages ranged between 23 and 49 years at a mean of 32.6 years. The average annual fuelwood demand per person in Teknaf, which was influenced by family size, was 1168 kg. Based on their fuelwood purchasing, collecting, and selling practices, local people were categorized into three groups: purchasers, collectors, and sellers. These groups evidenced significant differences in the duration of their residence in the area and in their annual incomes.

Keywords Fuelwood • Collection • Purchase • Selling • Consumption

S.M.A. Ullah (✉)
Department of Agricultural Extension Education, Faculty of Agriculture,
Bangladesh Agricultural University, Mymensingh 2202, Bangladesh
e-mail: asiksm@gmail.com

J. Tsuchiya
Department of Environmental Design, Faculty of Design, Kyushu University, Fukuoka, Japan
e-mail: j.tsuchiya.100@design.kyushu-u.ac.jp

M. Tani, M.A. Rahman (eds.), *Deforestation in the Teknaf Peninsula of Bangladesh*, DOI 10.1007/978-981-10-5475-4_7

7.1 Introduction

Rural communities in developing countries depend extensively on natural resources to sustain their livelihoods (Rennie and Singh 1996). Of these resources, forests continue to be integral to livelihood systems, especially of the poor, across the globe. An estimated 500 million people, who include some of the poorest rural inhabitants in the world, depend directly on forests for their livelihoods (Chao 2012). Forest products are the main sources of food, energy, medicine, animal feed, construction materials, furniture, agricultural implements, and utensils for many people around the world. Of these forest products, fuelwood, a non-timber forest product, which is the primary source of energy for many households, particularly those that are poor and rural, is the focus of this chapter. Its consumption as an energy source, constituting 9% of the total global energy consumption (Lauri et al. 2014), accounts for one of the main uses of forests and woodlands. Developing countries account for nearly 90% of the fuelwood that is produced and consumed worldwide (Dovie et al. 2004; Naughton-Treves et al. 2007). In South Asia alone, the demand for fuelwood is projected to reach 361.5 million cubic meters by 2020 (Arnold and Persson 2003), and in countries like Bangladesh, demands are apparently pressing up against limitations in supplies.

In Teknaf Upazila, which is the part of Bangladesh's coastal region, local people are dependent in multiple ways on forests, and especially on fuelwood, for their livings. Overdependence on forests is exerting pressure to an extent that threatens the future existence of the forest. Consequently, forest land and the resources they contain are steadily declining, day by day. The impact of fuelwood harvesting on forests is still a point of contention. Some studies have reported that there is a direct connection between fuelwood extraction and "severe deforestation" (e.g., Pang et al. 2013; Singh et al. 2010) or "forest degradation" (e.g., Ahrends et al. 2010; Cantarello et al. 2014; Moroni and Musk 2014; Orozumbekov et al. 2015; Ryan et al. 2012; Specht et al. 2015). On the contrary, other studies have suggested that fuelwood demands have limited impacts on forest cover (e.g., Hansfort and Mertz 2011; Shrestha et al. 2013). However, broad generalizations on fuelwood harvesting and deforestation are intrinsically misrepresentative. The effects of fuelwood collection on forests depend on various factors such as patterns of fuelwood supply and demand (Ghilardi et al. 2007; Wangchuk et al. 2014), type of usage as subsistence fuelwood or commercial charcoal (Naughton-Treves et al. 2007), vegetation responses to disturbances, changing species preference, extraction sites, and volumes extracted (He et al. 2009; Jagge and Shively 2014; Ruger et al. 2008).

Normally, harvesting fuelwood poses a small-scale disturbance to forests and is not considered a large threat to forests worldwide. However, the scenario in Teknaf is different. Here, fuelwood harvesting is the most prominent and visible subsistence activity conducted in forests. Moreover, the rural poor have access to very few fuel choices as alternatives to fuelwood obtained from forests. Fuelwood harvesting is therefore assumed to be one of the main causes of deforestation in Teknaf Upazila. Teknaf forests were declared as reserved forests in 1983 with the aim of protecting them and their resources. Subsequently, forests were officially reserved, and subsistence activities became illegal. However, inadequate regulation and overdependence of local people on forests have resulted in their continued exploitation. Apart from encroachment, overconsumption of forest resources in various forms has led to the current severely overexploited state of forests. Large-sized indigenous tree species have decreased at an alarming rate, and forest regeneration is being hampered by excessive collection of fuelwood. Generally, fuelwood is harvested in the form of twigs, dead branches, and leaves, but in Teknaf's forests, young saplings are being entirely uprooted and harvested for use as fuelwood. Moreover, frequent harvests from big trees hamper natural regeneration by reducing seed production. All of these practices are drastically affecting the forests. Consequently, fuelwood is an important component of sustainable forest management strategies in Teknaf.

Developing a comprehensive understanding of the situation regarding the demand, supply, and consumption of fuelwood in Teknaf has consistently been hampered by the lack of reliable information. Only a very small fraction of fuelwood production and consumption is recorded in Bangladesh. Given that fuelwood is mostly used by poor households, its consumption is seldom reported. It has, therefore, been difficult to assess the actual magnitude of fuelwood use and its impacts on forests and rural livelihoods (Arnold and Persson 2003). Specifically, in Teknaf, information regarding fuelwood consumption is very few. Uddin and Khan (2007) reported a figure for fuelwood consumption of 6 kg family^{-1} day^{-1}. An understanding of the demand, supply, and distribution of fuelwood is critical for effectively managing forests and ensuring a sustainable supply of forest resources in the future. This is because fuelwood is not only the major energy choice for cooking, but its collection and distribution also constitute the means of livelihood for many people in Teknaf. This chapter is organized into two parts. The first part focuses on the collection and distribution of fuelwood in Teknaf. The second part presents an analysis of the people associated with fuelwood collection and distribution. It also describes the fuelwood marketing system and supply chain. These analyses will help us to better understand fuelwood distribution and the key factors associated with fuelwood supply chain which, in turn, will facilitate the formulation of effective forest management strategies for protecting forests in the future.

7.2 Fuelwood Collection and Distribution

7.2.1 Approach

The study was conducted in Teknaf Upazila. Teknaf Upazila is situated in the southern coastal area of Bangladesh, and its population resides mainly around reserved forests. For most households, fuelwood is the only cooking energy option. Because of laxness in regulating access to reserved forests, and lack of other energy options, fuelwood harvesting and selling are widespread in Teknaf Upazila. Apart from the practice of self-collection by many households, a large proportion of fuelwood is distributed through large markets, locally termed bazars. There are four large bazars in Teknaf Upazila: Shamlapur, Nhilla, Teknaf, and Whykong (see Fig. 7.2). Fuelwood collectors bring bundles of fuelwood for sale at these bazars. A survey was conducted to obtain information on the weights and prices of fuelwood bundles. The fuelwood bundles were categorized into three groups based on their price: small bundles (1–50 BDT), medium bundles (50–180 BDT), and big bundles (above 180 BDT). Forty bundles within each category were weighed at each bazar. Several group discussions were also held with fuelwood distributors and sellers to acquire an understanding of the fuelwood supply chain. Data were collected during the period from September 2015 to December 2016.

7.2.2 Fuelwood Demand and Harvesting

Although forests in Teknaf Upazila are declared as reserved forests, the forests are widely accessed by local people. Regulation and supervision of Teknaf forest area are insufficient, which can prevent large-scale disturbances such as the harvesting of big trees and excessive encroachment, but it is not adequate to protect forests from being damaged by the small-scale subsistence activities of local people. Of these activities, fuelwood collection is the most visible and prominent. Poor communities in Teknaf have no alternative to fuelwood as an energy source for their cooking requirements. Moreover, easy access to forests encourages people to harvest fuelwood free of cost. Almost every household in Teknaf uses fuelwood for cooking. Those households that are comparatively solvent can purchase fuelwood from local sellers or at the bazars, whereas members of relatively poor households enter the forests themselves to collect fuelwood. Apart from using fuelwood for cooking, many local people harvest, supply, and sell it as a means of livelihood. Fuelwood is generally collected throughout the year. However, during the winter season (from November to February), fuelwood harvesting reaches a peak. During this period, many households harvest and stack fuelwood for the entire year. Local people harvest fuelwood and carry it as headloads from the forests. They usually have to walk for 1–3 h in the forests to select a specific site for harvesting fuelwood. After harvesting the fuelwood, they make up bundles weighing from 5 to 30 kg. The size of a bundle depends on the type of fuelwood and the amount that the harvester can carry. In Teknaf, there are few trees in the reserved forests. Mostly, saplings and the twigs and branches of big trees are harvested as fuelwood. Local people, gathering fuelwood, enter and leave the forests along particular trails. We identified 111 trails (Fig. 7.1) around the reserved forests in Teknaf. These trails are widely used by local people all year round.

The actual demand for fuelwood in Teknaf Upazila is not clear because of the lack of data. In 2015, we conducted a survey on fuelwood consumption in a village called Jahajpura in Teknaf. For this study, we collected data through a household survey. A total of 42 out of 470 households in Jahajpura were interviewed using a

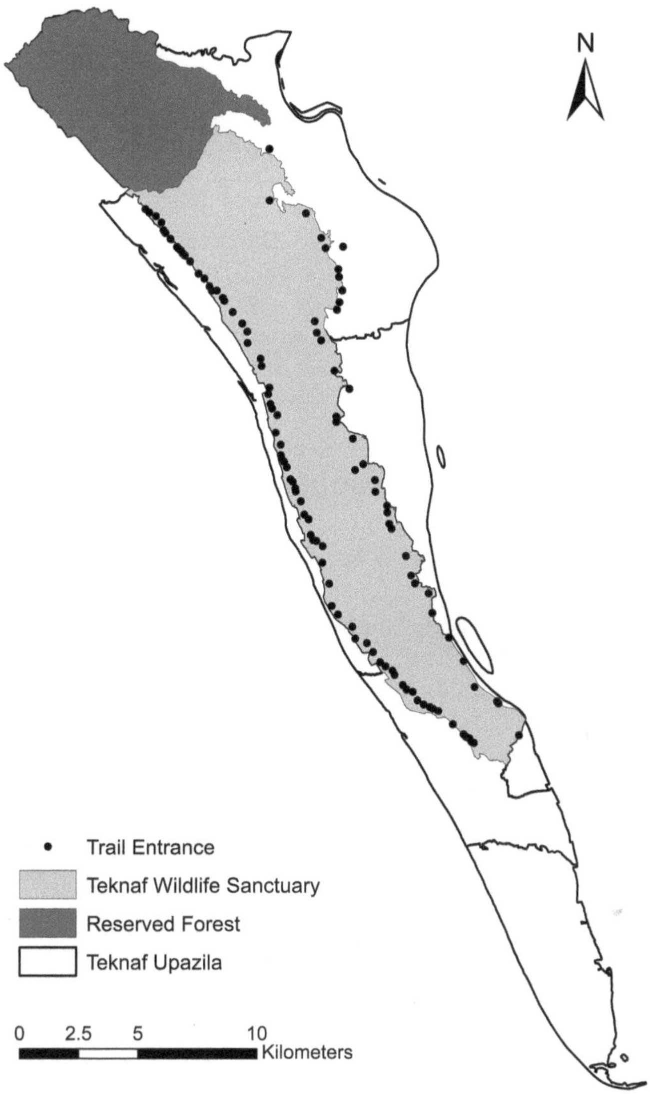

Fig. 7.1 Trail entry points in Teknaf forest

structured questionnaire. To determine which households to survey, all of the households in the village were categorized into three groups based on family size. These groups comprised small families (up to five members), medium-sized families (6–10 members), and large families (more than 10 members). Stratified random sampling was used to select households within these groups. Fourteen households were selected from each of the three categories, resulting in a total sample size of 42 households. For estimating fuelwood consumption, we asked members of the sampled households to show us their stacked fuelwood bundles used for cooking. Out of these bundles, and based on the family size, three to five bundles were weighed and separated from the rest of the fuelwood. The respondents were then requested to cook for the household using the weighed bundles for the next 3 days. After this period, the remaining bundles were weighed again. After physically weighing fuelwood bundle, the weight was divided by family size to determine per capita consumption. The overall mean of daily per capita fuelwood consumption was calculated to be 3.2 kg (1168 kg/person/year). The household fuelwood consumption increased with an increase in family size. The quantities of fuelwood consumed by small, medium-sized, and large families were 13 kg, 21 kg, and 34 kg, respectively. However, with the upsurge in family size in the area, per capita fuelwood consumption has decreased. Daily fuelwood consumption per household ranged from a high value of 37.25 kg to a low value of 6.92 kg.

7.2.3 Fuelwood Marketing and Distribution

We found that a major proportion of fuelwood in Teknaf was harvested by households for their own consumption and did not pass through marketing and distribution channels. Apart from self-collected fuelwood, people also purchased fuelwood from local collectors and at the big bazars. Although there is no formal fuelwood marketing sector, a numerically small but highly visible private sector exists in Teknaf for marketing and distributing fuelwood. Their main activities centered on the four big bazars in Teknaf Upazila, namely, Shamlapur, Nhilla, Teknaf, and Whykong (see Fig. 7.2). Table 7.1 shows the price per kg of fuelwood at these four bazars.

The price of fuelwood was lowest at Shamlapur Bazar and highest at Nhilla Bazar. These price differences can be attributed to different sources and transportation costs of fuelwood. Sources of fuelwood sold at the different bazars were as follows. Fuelwood sold at Shamlapur Bazar was sourced from forests located near Shamlapur and Monkhali. Fuelwood sold at the Teknaf Bazar was sourced from forests located near Shamlapur, Dumdumia, and Mochani. Fuelwood sold at Whykong Bazar was sourced from forests located near Palonkhali, Monkhali, and Whykong. Lastly, fuelwood sold at Nhilla Bazar was sourced from forests located near Palonkhali and Kombon. To acquire a better understanding of the fuelwood marketing and distribution system, we held several group discussions with fuelwood vendors engaged in the marketing and distribution of fuelwood. Among the 57 vendors with whom we held group discussions, 35 identified fuelwood sales as their main income source. We identified three types of fuelwood vendors in Teknaf: (1) those collecting fuelwood from forests and selling it at the market or to neighboring households or other vendors, (2) those buying fuelwood from small vendors and selling it at the market, and (3) those buying fuelwood from small vendors at the big bazars and also selling this fuelwood at the same bazar.

In Teknaf, the distribution channel begins with collectors, who mainly comprise the first type of fuelwood vendors (Table 7.2). They either sell to neighboring households or to other vendors. They also come to the market to sell fuelwood. Normally,

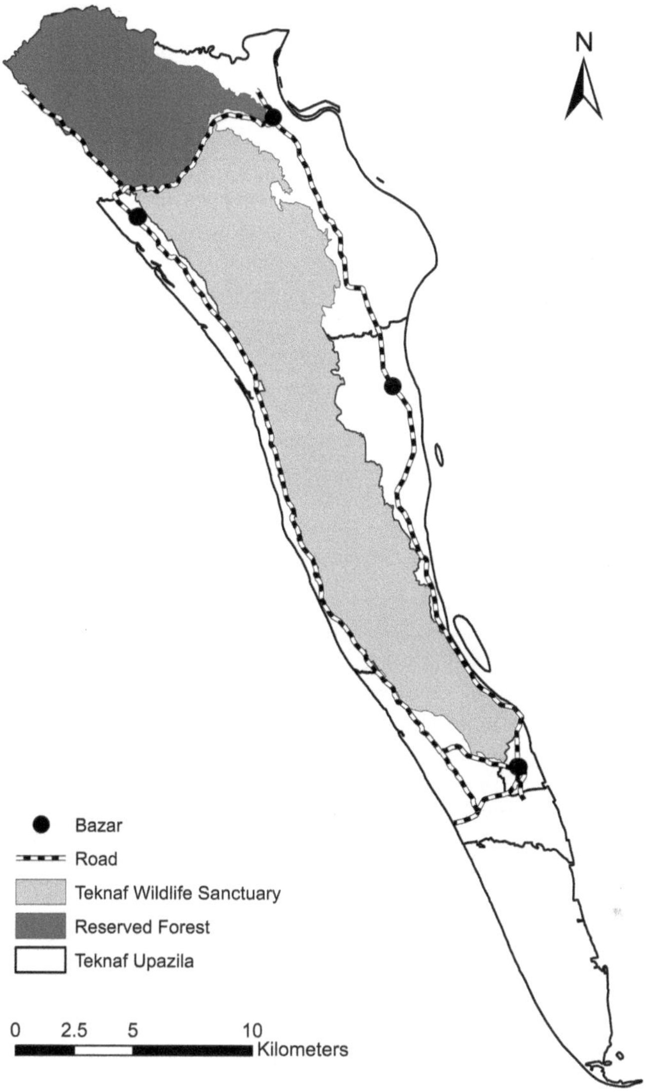

Fig. 7.2 Map showing different bazars for fuelwood in Teknaf

Table 7.1 Market price (BDT bundle^{-1}) of fuelwood in different bazars of Teknaf Upazila

Bundle size	Shamlapur Bazar	Whykong Bazar	Nhilla Bazar	Teknaf Bazar
Small bundle	4.66	4.79	9.80	15
Medium bundle	6.88	6.63	8.37	10.85
Large bundle	6.57	8.11	9.11	7.71
Total	**6.35**	**7.04**	**8.87**	**8.56**

Table 7.2 Core results from the fuelwood vendor survey

Sample sizes	Overall n = 57: Shamlapur Bazar, 23; Teknaf Bazar, 17; Nhilla bazar, 10; Whykong bazar, 7
Gender	63% were male and 37% were female
Age	Ranged between 23 and 49 with a mean of 32.6 years
Location of operation	57% of vendors come from the nearby villages and sell in the big bazars
	26% of vendors came from other unions
	17% comes from other Upazilas
Scale of operation	70% of vendors presented more than 100 kg fuelwood on a market day
Frequency of operation	19 full-time vendors (5–7 days/week), 28 part-time (1–4 days/week), and 10 of the vendors were irregular supplier
Source of fuelwood	52% of vendors collect fuelwood from protected forest area, 28% buy from other collectors and transport to market for selling, and 23% buy from small vendors and sell to others in the big markets
Transportation	Pickup lorry, 24 vendors; small three wheelers, 17 vendors; headload, 16 vendors
Transportation cost	A big bundle costs around 10 BDT and small ones 5 BDT, small three wheelers' rent ranges from 20 to 150 based on distance and number of bundles

they sell fuelwood in small amounts. Fuelwood vendors belonging to the second group collect fuelwood from small vendors and transport it to the big bazars. These vendors sell fuelwood in large quantities. Vendors belonging to the third group purchase fuelwood from other vendors and sell it in large quantities. These vendors sell fuelwood almost every day of the week. Fuelwood prices at the bazars rise because of transportation costs. The high selling prices of fuelwood at the bazar motivate small vendors to come to the big bazars to sell fuelwood.

7.3 People Associated with Fuelwood Collection, Purchase, and Sales

7.3.1 Approach

This section of the chapter identifies the socioeconomic characteristics of local people involved in the collection, purchasing, and selling of fuelwood. As discussed in Chap. 2, data were collected through household surveys conducted from December 2015 to May 2016. A structured questionnaire was used to collect data from households located within the six unions. The data collection process covered all available households within each union. Data obtained from a total of 5769 households were analyzed. Based on their activities relating to the collection and sale of fuelwood, households were categorized into three groups: purchasers, collectors, and sellers. The group of purchasers comprised households that mainly purchased fuelwood for household use, but whose members did not themselves go to the forest to collect fuelwood. The collector group comprised households that collected

Table 7.3 Distribution of people according to fuelwood collection, union, and living in and out of the forest

Union category		Baharchhara	Nhilla	Sabrang	Teknaf town	Teknaf Sadar	Whykong	Total
Purchaser	Outside forest	335	638	929	320	580	620	3691
	Inside forest	74	12	0	47	41	23	
Collector	Outside forest	296	376	217	8	347	368	1979
	Inside forest	135	21	0	4	86	121	
Seller	Outside forest	6	33	1	1	4	67	171
	Inside forest	10	6	0	1	10	32	
Total		856	1086	1147	381	1068	1231	5769

fuelwood in the forest for their own use but did not sell fuelwood. The last group comprised sellers who both collected and sold fuelwood either in their own localities or at markets. Table 7.3 shows the distribution of households according to these three categories in the six unions and the location of the households either in or outside forests. Figure 7.3 shows the settlement patterns of the three categories of households. The groupings of households were made based on the engagement of household heads in purchasing, collecting, or selling fuelwood. The age, education, family size, duration of residence, annual income, occupation, and capital generation of a household head was used to describe the characteristic features of each household. Averages, percentages, and other statistics were used for describing the characteristics of household respondents. The presentation of data in tables and graphs improved the clarity of the analysis. To demonstrate variability among the three groups, one-way ANOVA test was carried out. The Windows SPSS 16.0 statistics program was used to perform these tests.

7.3.2 A Socioeconomic Profile of Households Associated with Fuelwood

The study revealed the socioeconomic characteristics of the three groups and compared differences between them. Table 7.4 shows comparative data relating to the socioeconomic characteristics of fuelwood purchasers, collectors, and sellers. Whereas the mean values for education, duration of residence, and annual incomes showed variations, the means for age and family size did not show much variation. From Table 7.4, it is apparent that incomes and education levels were higher for the purchaser group than for the other groups. Incomes and education levels were lowest for the seller group.

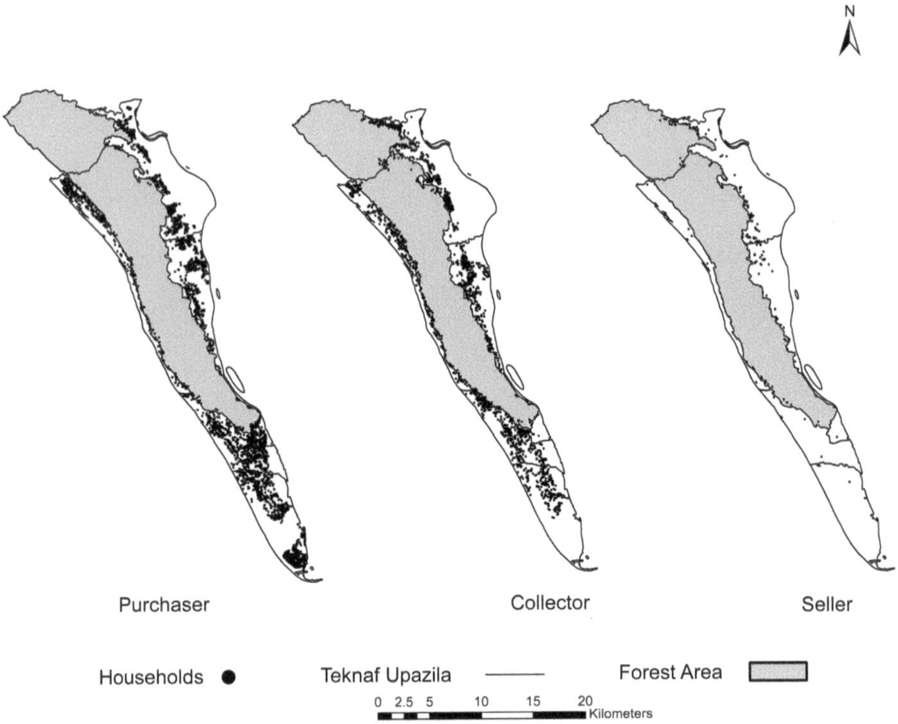

Fig. 7.3 Settlement patterns of purchaser, collector, and seller of fuelwood

Table 7.4 Socioeconomic characteristics of the purchaser, collector, and seller in Teknaf

Socioeconomic characteristics	Purchaser	Collector	Seller	Total
Age (years)	39.93	40.78	40.34	40.23
Education (schooling year)	2.18	1.66	0.94	1.97
Family size (family members)	6.15	6.17	5.72	6.14
Length of residence (years)	8	18	13	13.36
Annual income (000 BDT)	211	169	163	195

Age The mean age of household heads in the study area was 40 years. There were few age-related differences between categories and groups, namely, purchasers, collectors, and sellers in and outside forests in the unions of Baharchhara, Nhilla, Sabrang, Teknaf town, Teknaf Sadar, and Whykong. Figure 7.4 also shows that there were very few variations between the three groups. The mean age of household heads in the seller group in Teknaf town was highest at over 60 years.

Education Education levels were determined by the number of years of schooling of household heads. The mean number of years of schooling for all groups was 1.97 years. The mean numbers of years of schooling for purchasers, collectors, and sellers were 2.1, 1.6, and 0.9 years, respectively. The average education level within

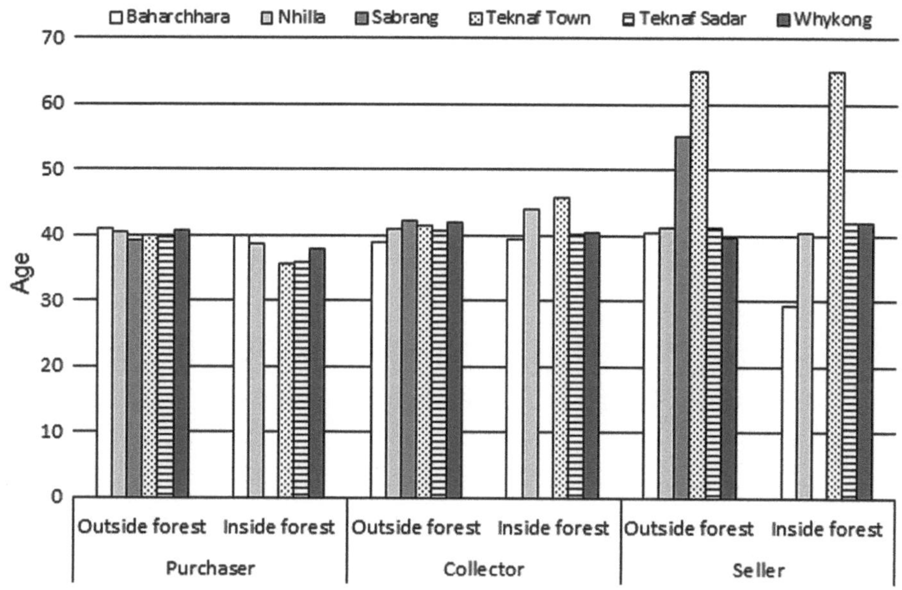

Fig. 7.4 Distribution of age according to their category

the purchaser group was higher than the average education levels of household heads in the collector and seller groups in all unions apart from Sabrang. The mean number of years of schooling within the seller group was less than a year. The mean number of schooling years in Baharchhara, Nhilla, Sabrang, Teknaf town, Teknaf Sadar, and Whykong were 1.7, 2.5, 1.2, 2.0, 2.0, and 2.3 years, respectively. Differences between the three categories were significant (F 32.120 and Sig .000), as were differences among sellers (F 13.23 and Sig .003) and collectors (F 25.16 and Sig .007), depending on whether they lived in or outside forests. Whereas education levels were almost the same for purchaser groups, regardless of whether they lived in or outside forests, there were differences between households belonging to the collector and seller groups living within and outside forest boundaries. Figure 7.5 shows that households belonging to both the collector and seller categories who lived in forests had lower education levels than those living outside forests.

Family Size Family size was measured according to the number of people living together and cooking in the same kitchen. In general, large families are predominant in rural areas of Bangladesh, and the Teknaf Peninsula is no exception. In this area, the mean family size was 6.14. There were no significant differences between the three groups based on family size, but significant differences were found in relation to unions (F 3.005 and Sig 0.010). Figure 7.6 shows that the mean number of family members in households of fuelwood sellers living in the forests in Teknaf town was the highest, whereas the mean family size of fuelwood sellers living in forests in Baharchhara was the lowest.

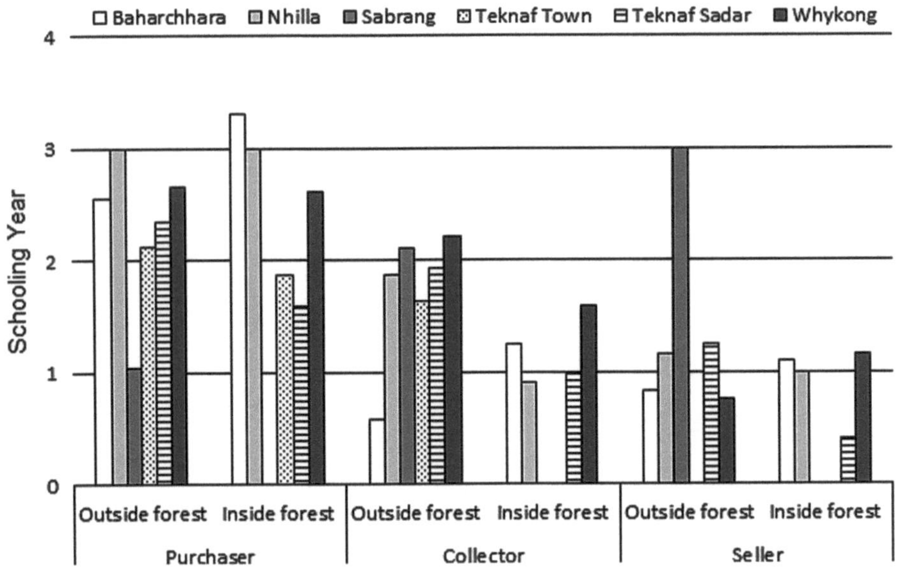

Fig. 7.5 Distribution of schooling year according to their category

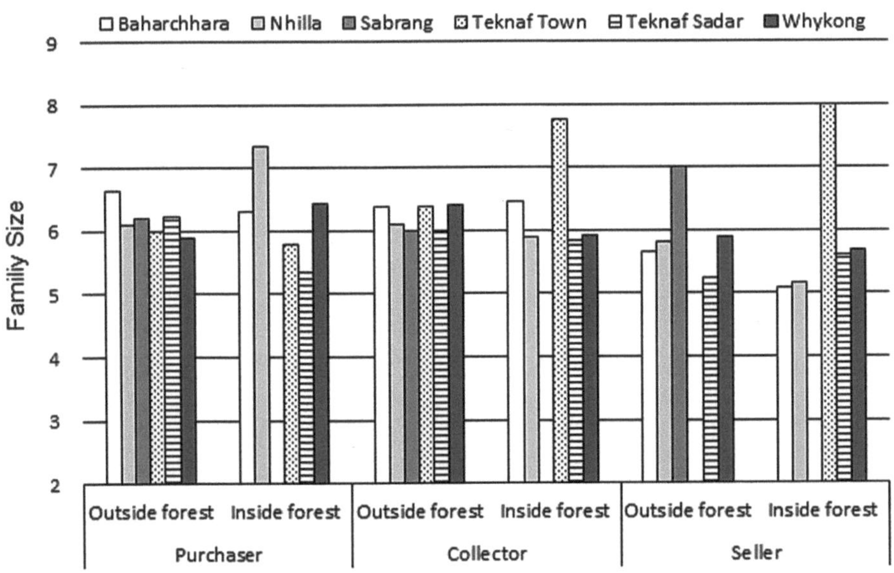

Fig. 7.6 Distribution of family size according to their category

Duration of Residence Duration of residence was measured by the duration of a family's residence in its current house. The mean duration of residence was about 13 years. The mean durations of residence of purchasers, collectors, and sellers were 8 years, 18 years, and 13 years, respectively. There were significant differences in the three categories (F 5.028 and Sig 0.007). Significant differences were

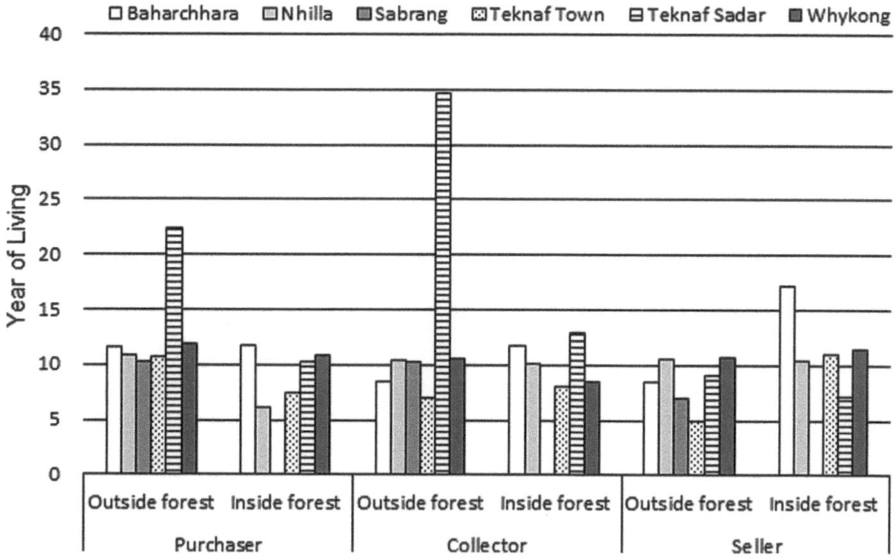

Fig. 7.7 Distribution of year of living according to their category

also found for households living in or outside forests among groups that purchased (F 26.49 and Sig .000) and collected fuelwood (F 23.76 and Sig .001). Figure 7.7 shows that the durations of residence of purchasers and collectors in Teknaf Sadar union were longer than that of other groups in different unions.

Occupations and Capital Generation Capital generation was measured by the percentage of total annual incomes obtained from different livelihood sources, namely, farming (excluding *pan* cultivation), *pan* cultivation, fishing, labor, business, and employment abroad. The highest proportion of household incomes in the group of fuelwood purchasers living outside forests was derived from business (24%), and the lowest proportion was derived from farming (7%). For fuelwood collectors living outside forests, labor (26%), farming (14%), and fishing (13%) were the major income sources. On the other hand, the main income sources of fuelwood sellers living outside forests were labor (71%) and other sources (13%). The highest proportion of household incomes of fuelwood purchasers living in forests was derived from other sources (26%), and the lowest proportion was derived from farming (2%). For fuelwood collectors living in forests, labor (29%), *pan* farming (18%), and farming (13%) were the major income sources. For fuelwood sellers living in forests, labor (56%) and farming (12%) accounted for the highest proportions of incomes (see Fig. 7.8).

Annual Incomes Households' annual incomes were calculated by combining their separate incomes derived from different livelihood sources such as farming, fishing, labor, employment abroad, and business. There were significant differences between groups of fuelwood purchasers, collectors, and sellers (F 31.532 and Sig .000) according to whether households were living in or outside forests (purchasers, F 15.37 and Sig 0.014; collectors, F 21.33 and Sig 0.002; and sellers, F 30.57 and

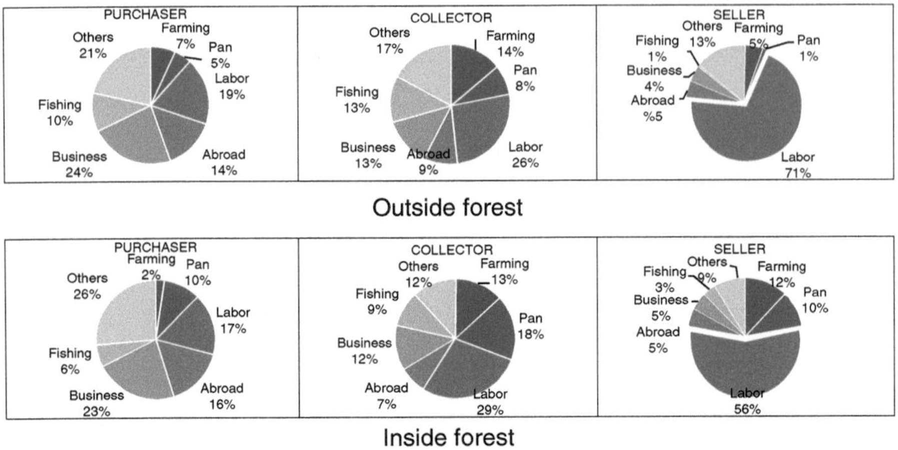

Fig. 7.8 Different income sources of three groups

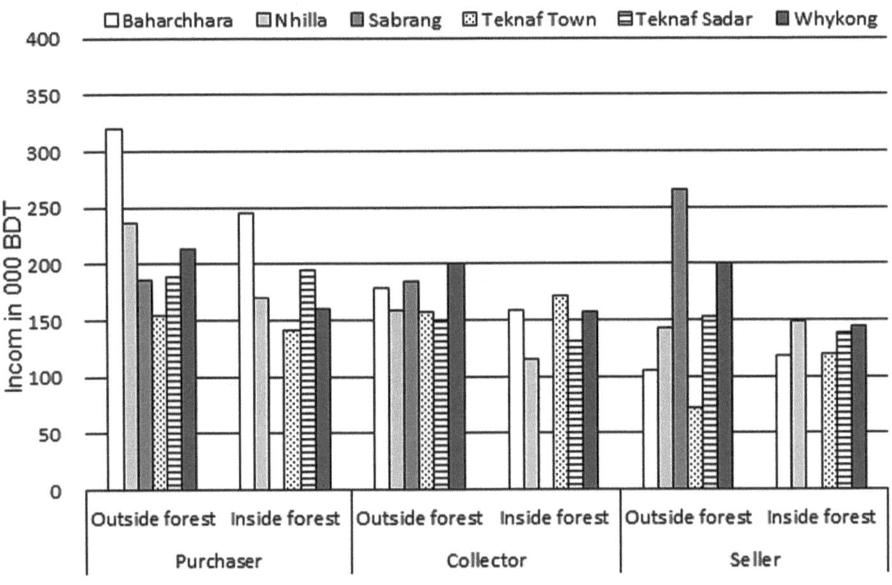

Fig. 7.9 Mean annual income of purchaser, seller, and collector (000 BDT)

Sig 0.000). Table 7.4 shows mean annual incomes of each of the three groups from different sources. For each group, the mean was calculated by dividing the total income obtained from a specific income source by the number of households earning from that source. The mean total annual income for all households was 195,000 BDT (Fig. 7.9). Of the households living in and outside forests, those living outside forests were more affluent than those living in forests. Moreover, the incomes of fuelwood purchasers were higher than those of collectors and sellers.

7.4 Conclusion

The average demand for fuelwood in Teknaf is 1168 kg person^{-1} year^{-1}. Fuelwood is not just the main source of cooking energy; its supply and distribution also provide a means of livelihood for many poor people. Fuelwood harvesting is an important factor for deforestation in Teknaf Upazila. Thus, sustainable use of fuelwood can help to preserve forests for the future. Whereas the use of alternative energy sources for cooking would divert people from using fuelwood, alternative fuel choices such as LP gas are costly, and their adoption by people in Teknaf will take time. Consequently, fuelwood is an important factor to be considered when formulating forest protection strategies. Proper understanding on the demand, supply, and marketing of fuelwood will provide a clear prospect to the total fuelwood scenario in Teknaf, which is required for achieving sustainable fuelwood consumption from the local forest areas. Based on their collection and sales of fuelwood, local people were categorized into three categories: purchasers, collectors, and sellers. The purchaser group evidenced better incomes, longer duration of residence, and more education compared with other groups. Significant differences were found in annual incomes, education, and duration of residence among the groups. There were significant differences in income, education, and duration of residence within each union, while differences in income and duration of residence were significant between households living in and outside forests.

References

Ahrends A, Burgess ND, Milledge SAH, Bulling MT, Fisher B, Smart JCR, Clarke GP, Mhoro BE, Lewis SL (2010) Predictable waves of sequential forest degradation and biodiversity loss spreading from an African city. Proc Natl Acad Sci USA 107(33):14556–14561

Arnold M, Persson R (2003) Reassessing the fuel wood situation in developing countries. Int For Rev 5(4):379

Cantarello E, Lovegrove A, Orozumbekov A, Birch J, Brouwers N, Newton AC (2014) Human impacts on forest biodiversity in protected walnut-fruit forests in Kyrgyzstan. J Sustain For 33(5):454–481

Chao S. 2012 Forest peoples: numbers across the world. In: Forest Peoples Programme, Moreton-in-Marsh UK

Dovie DBK, Witkowski ETF, Shackelton CM (2004) The fuelwood crisis in Southern Africa – relating fuelwood use to livelihoods in a rural village. GeoJournal 60(2):123–133

Ghilardi A, Guerrero G, Masera O (2007) Spatial analysis of residential fuelwood supply and demand patterns in Mexico using the WISDOM approach. Biomass Bioenergy 31(7):475–491

Hansfort SL, Mertz O (2011) Challenging the woodfuel crisis in West African woodlands. Hum Ecol 39(5):583–595

He G, Chen XD, Bearer S, Colunga M, Mertig A, An L, Zhou S, Linderman M, Ouyang ZY, Gage S, Li S, Liu JG (2009) Spatial and temporal patterns of fuelwood collection in Wolong Nature Reserve: implications for panda conservation. Landsc Urban Plan 93(1):92

Jagge P, Shively G (2014) Land use change, fuel use and respiratory health in Uganda. Energy Policy 67:713–726

Lauri P, Havlı KP, Kindermann G, Forsell N, Böttcher H, Obersteiner M (2014) Woody biomass energy potential in 2050. Energy Policy 66:19–31

Moroni MT, Musk RA (2014) Domestic timber harvesting affects wood quantities in Tasmanian dry eucalypt forests. Aust For 77(2):86–91

Naughton-Treves L, Kammen DM, Chapman C (2007) Burning biodiversity: woody biomass use by commercial and subsistence groups in western Uganda's forests. Biol Conserv 134:232–241

Orozumbekov A, Cantarello E, Newton AC (2015) Status, distribution and use of threatened tree species in the walnut-fruit forests of Kyrgyzstan. For Trees Livelihoods 24(1):1–17

Pang C, Yu H, He J, Xu J (2013) Deforestation and changes in landscape patterns from 1979 to 2006 in Suan County, DPR Korea. Forests 4(4):968–983

Rennie J, Singh N (1996) Participatory research for sustainable livelihoods: a guidebook for field projects. International Institute for Sustainable Development, Winnipeg

Ruger N, Williams-Linera G, Kissling WD, Huth A (2008) Long-term impacts of fuelwood extraction on a tropical montane cloud forest. Ecosystems 11(6):868–881

Ryan CM, Hill T, Woollen E, Ghee C, Mitchard E, Cassells G, Grace J, Woodhouse IH, Williams M (2012) Quantifying small-scale deforestation and forest degradation in African woodlands using radar imagery. Glob Change Biol 18(1):243–257

Shrestha S, Karky BS, Gurung A, Bista R, Vetaas OR (2013) Assessment of carbon balance in community forests in Dolakha, Nepal. Small-Scale For 12(4):507–517

Singh G, Rawat GS, Verma D (2010) Comparative study of fuelwood consumption by villagers and seasonal Dhaba owners in the tourist affected regions of Garhwal Himalaya, India. Energy Policy 38(4):1895–1899

Specht MJ, Ribeiro PSR, Albuqueque UP, Tabarelli M, Melo Felipe PL (2015) Burning biodiversity: fuelwood harvesting causes forest degradation in human-dominated tropical landscapes. Glob Ecol Conserv 3:200–209

Uddin MS, Khan MASA (2007) Comparing the impacts of local people and Rohingya refugees on Teknaf game reserve, Bangladesh. In: Fox J, Bushley B, Dutt S, Quazi SA (eds) Making conservation work: linking rural livelihoods & protected area management in Bangladesh. East-West Center and Nishorgo Program of the Bangladesh Forest Department, Dhaka, pp 149–175

Wangchuk S, Siebert S, Belsky J (2014) Fuelwood use and availability in Bhutan: implications for national policy and local forest management. Hum Ecol 42(1):127–135

Chapter 8
Forest Dwellers and Encroachers in Teknaf

Masakazu Tani

Abstract This chapter begins by providing an overview of the population distribution within as well as outside of the restricted forest area in the study region. Next, it presents the findings of a case study on the socioeconomic attributes and activities of residents by ethnic groups in a village consisting of settlements within and outside of the Teknaf Wildlife Sanctuary. Although reserved forests are legally closed to private activities, with the exception of those of certain categories of people who have special rights to live in the area, about half of these forest dwellers are "illegal" encroachers. Whereas Bengalis, who legally reside within the reserved forest area, earn comparable incomes to those of people living outside of the reserved forest area, the incomes of encroachers and minority ethnic groups are significantly lower. Many of the encroachers moved relatively recently into this restricted area, and the inflow of settlers does not appear to be lessening. Because the encroachers' primary means of subsistence entails farming that requires the clearing of existing forests, the increase in their numbers appears to constitute a major threat to forest regeneration in this area.

Keywords Encroacher • Ethnicity • Forest villagers • Settlement • Teknaf Wildlife Sanctuary

8.1 Introduction

Gibson et al. (2000) emphasized the important role of local communities in conserving forests. They argued that national governments in developing countries often lack sufficient resources to enforce forest preservation laws. Compared with national governments that can only have an indirect impact on forest-related activity, "local communities live with forests, are primary users of forest products, and create [their own] rules that significantly affect forest condition" (Gibson et al.

M. Tani (✉)
Department of Environmental Design, Faculty of Design, Kyushu University, Fukuoka, Japan
e-mail: tani@design.kyushu-u.ac.jp

M. Tani, M.A. Rahman (eds.), *Deforestation in the Teknaf Peninsula of Bangladesh*, DOI 10.1007/978-981-10-5475-4_8

2000, p. 3). Therefore, a better understanding of the characteristics of local populations that play a key role in preserving or destroying forests is critical.

This chapter examines the characteristics of the population residing in the Teknaf Wildlife Sanctuary (TWS) and compares their characteristics with those of the population residing near the sanctuary. As noted in previous chapters, the TWS is a reserved forest area and is therefore closed, in principle, to human activities. However, this study found that a sizable population resides within its boundaries, with not all residents in the restricted area being illegal encroachers. Because these forest dwellers significantly affect the state of the forest, this chapter begins by presenting data on the population distribution in the restricted TWS and adjacent reserved forest area based on the results of a household location survey. It then presents the findings of a detailed village case study conducted to analyze and compare the socioeconomic attributes and livelihood activities of people living in the restricted forest areas with those of the neighboring local population located outside of the restricted area. This analysis enabled us to assess the role of local people in the area's deforestation.

8.2 The Local Population Living in the Restricted Forest Area

To conduct the project's baseline survey, all households (57,404) within Teknaf Upazila were visited by project staff, and their locations were recorded using GPS devices (see Sect. 2.5.2 of Chap. 2). Ten percent of the households (5769) were sampled, and demographic and socioeconomic characteristics were recorded using a questionnaire.

Households located within the boundaries of the restricted forest area were identified using the recorded GPS locations (longitude–latitude) and forest boundaries indicated on cadastral maps. These cadastral maps showed the boundaries of privately owned land plots as well as those of public lands, including reserved forests. The cadastral maps were first scanned and geo-referenced on to a GIS system, and the locations of households were overlaid on the forest boundaries to distinguish households that were located within and outside of the restricted area (Fig. 8.1). Table 8.1 presents the union-wise distribution of households according to their locational status within and outside of the reserved forest area.

Table 8.1 shows that of a total of 57,404 households, 5195 households (9.0%) were located in the restricted area. Although no population count was attempted, the population inside the restricted forest area was estimated to be approximately 32,000, based on a mean household size of 6.1 persons that was calculated using data obtained from a 10% sample of all households. There were more households located within the reserved forest area on the western side of the peninsula than the eastern side. Twenty-seven percent of all households in Baharchhara Union were located inside the TWS. The area of this union comprises a very narrow strip of land stretching between the coastal line and hill ranges. The fact that plains are very

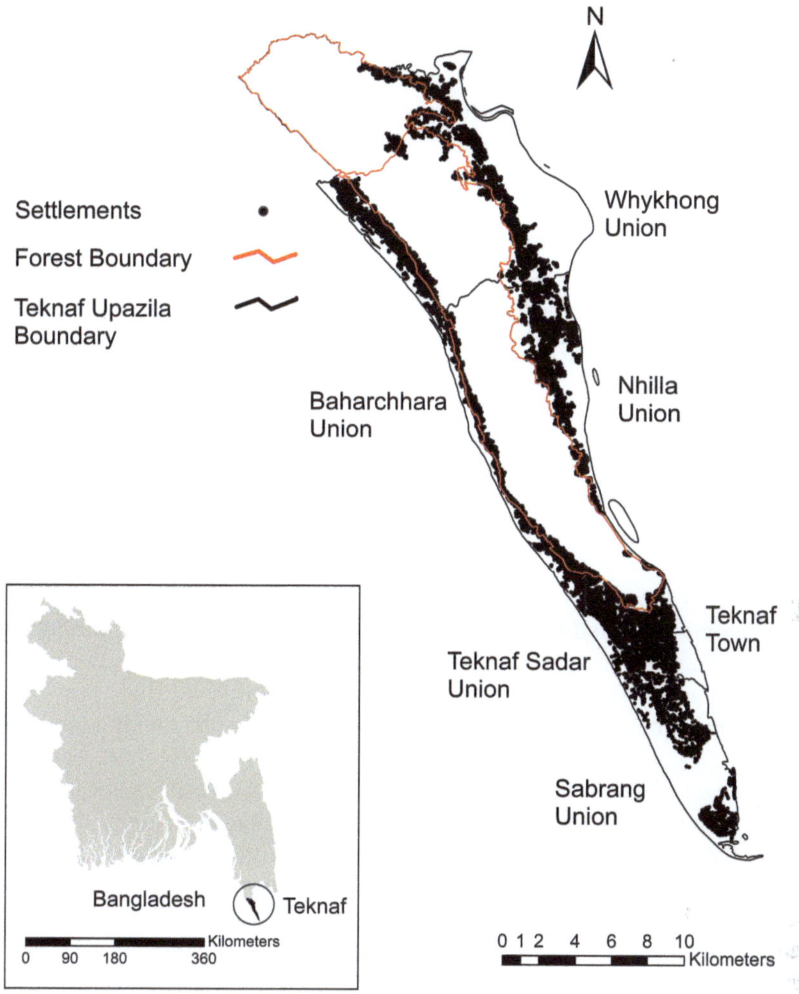

Fig. 8.1 Distribution of households in Teknaf Upazila. Boundaries of reserved forests are shown in *red* and those unions in *black*

Table 8.1 List of recorded categories in the survey

Household attributes	
	Location, religion, ethnicity, habitation history, and household income by farming, betel leaf (*pan*) farming, and other income-generating activities
Attributes of household members	
	Age, sex, education, and birth place

limited in Baharchhara may partly account for the high proportion of the households belonging to this union that are located within the TWS. By contrast, relatively fewer households in Hnilla and Whykong Unions, which are on the eastern side of the peninsula, are located inside the reserved forest area. Plain areas are comparatively wider on this side of the peninsula than those on the western side. Some of the households belonging to Teknaf Town and Teknaf Sadar Union, which are located south of the TWS, also reside inside the TWS, but their proportion, in relation to the total population, approximated to a figure halfway between the proportions living in this restricted area along the eastern and western sides of the peninsula. Because Sabrang Union does not share a border with the TWS, there were no households from this union located within the TWS.

8.3 Case Study

8.3.1 Settings

As revealed by the results of the baseline survey, there is a sizable resident population within the restricted forest area. Although reserved forests are closed, in principle, to private human activities, certain groups of people are legally permitted to live in these areas. This section presents the findings of a case study that examined the socioeconomic characteristics and legal status of dwellers within the restricted forest area.

The village of South Shilkhali was selected for this case study for two reasons. The first relates to the location of the village on the western side of the peninsula (see Fig. 8.2), which has a high population density within the restricted forest area. The second reason for selecting this village was because it contained a few smaller, and more or less discrete, settlements known locally as *para* that are located both within and outside of the forest boundaries, thereby facilitating a comparison between the two locational statuses. Of these *paras* within the village, Naya, South Chakma, and Math are fully contained within the boundaries of the TWS's reserved area, whereas Kader *para* is located outside of the sanctuary's boundaries (see Fig. 8.3). Although this village also includes other settlements, only the above four settlements were surveyed.

Questionnaire-based surveys were conducted in September 2012 in the *paras* within the TWS, namely, Naya, South Chakma, and Math, and a further survey was conducted in Kader *para* outside the TWS in September 2013. A total of 171 households belonging to the three *paras* within the TWS were recorded, and 85 households were recorded outside of the TWS in Kader *para*. No sampling technique was used to select particular households within each *para*, but the responses to the questionnaire items of all of the available households in the surveyed *paras* were recorded. These items are listed in Table 8.1.

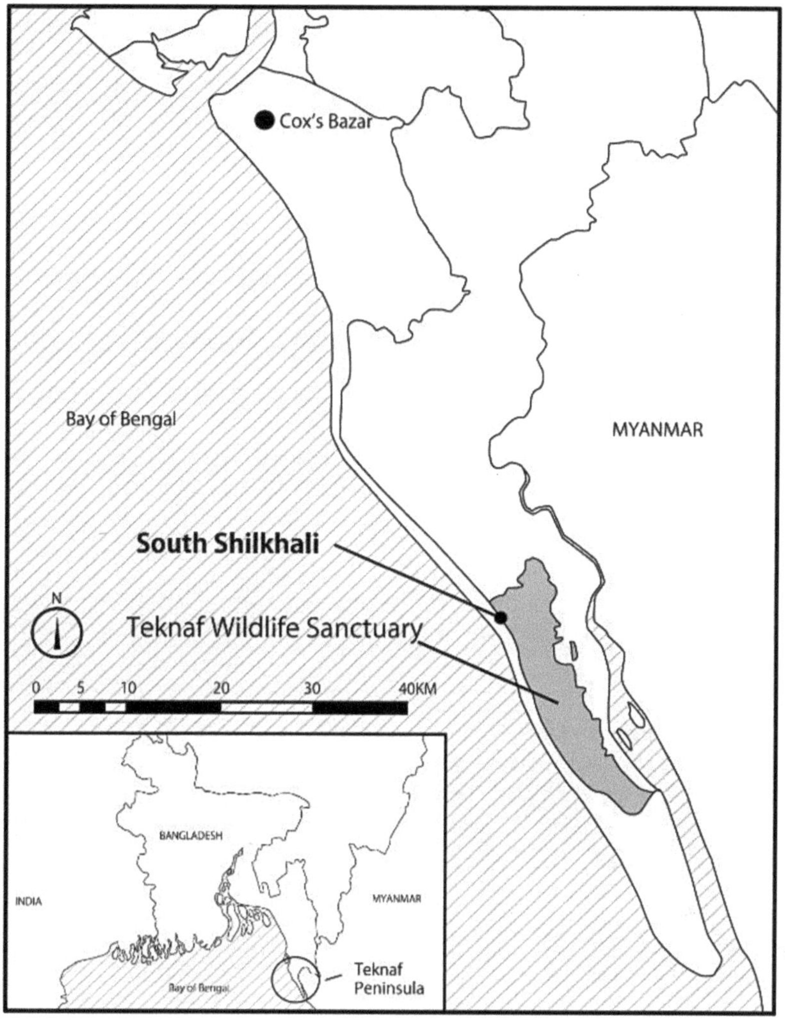

Fig. 8.2 Map of the Teknaf Peninsula and the location of South Shilkhali

8.3.2 Ethnicity, Encroachers, and Legitimate Residents

Three ethnic groups were represented among the recorded households in South Shilkhali. Bengalis, who predominate in Bangladesh, comprised the largest ethnic group. There were 139 households located in the three *paras* within the TWS, and 85 located in Kader *para* outside the TWS. The only ethnic group recorded outside the TWS was Bengali. Two minority groups were represented inside the TWS: the Chakma and Rohingya. There were 26 Chakma households and 6 Rohingya households within the sanctuary (see Table 8.2). The Chakma are a Buddhist Tibeto-Burman group of indigenous ethnicity. The largest Chakma population resides in

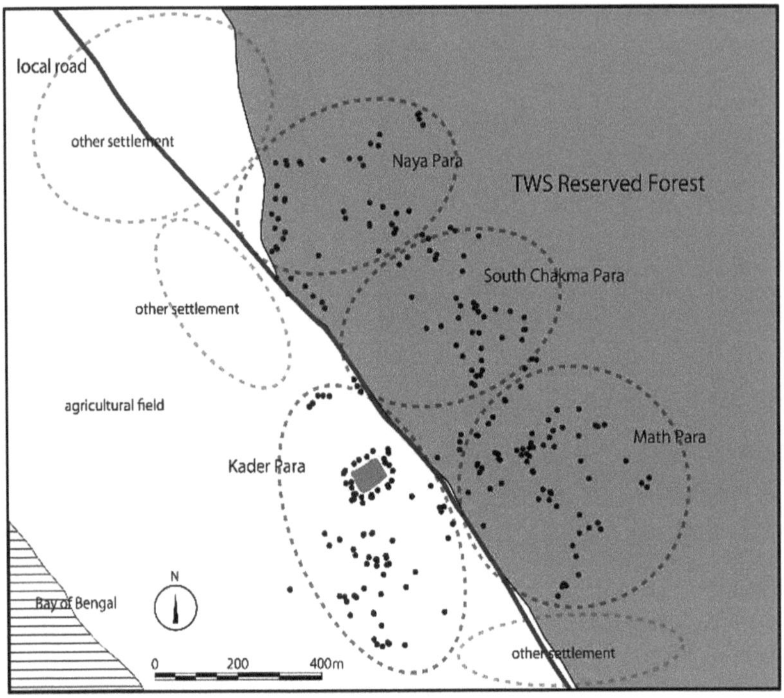

Fig. 8.3 Distribution of paras in South Shilkhali and the area of the TWS. *Dots* represent houses in the surveyed area. Because villages and paras in this area are groups of houses without definite boundaries, the borders of paras are indicated by *dotted lines*

Table 8.2 The number of households by ethnic groups inside and outside the TWS reserved forest in four settlements in South Shilkhali

Location	Ethnicity	No. of households
Inside the TWS	Bengali	139
	Chakma	60
	Rohingya	6
Outside the TWS	Bengali	85

the Chittagong Hill Tract, and their distribution extends into the surrounding areas, including the Teknaf Peninsula. The Rohingya are a Muslim group residing in Myanmar. However, during the last few decades, significant numbers of Rohingya refugees fled to Bangladesh to escape sociopolitical oppression.

The legal status of dwellers requires further explanation. Although the reserved forest is basically closed to private activities, there are several categories of people who are legally entitled to live in the reserved forest. The first category is forest villagers, who are registered with the Forest Department (FD) and are permitted to live in the area in exchange for performing forest management duties (Akhter et al. 2009; IPAC 2009). Each household belonging to the forest villager category is allotted a two-*kani* (0.81 ha) plot by the FD. Forest villager status can be inherited but only by one descendant of a deceased forest villager. Strictly speaking, those who do not inherit this status do not have the right to live in the reserved forest area after the land-owning parent dies. However, descendants of these non-heirs tend to remain in the forest and are therefore included in the forest villager category in this study.

Families who started living in reserved forests before the Forest Act was enacted in 1927 constitute a second category of legal residents in relation to reserved forests. If a family was resident in the current reserved area prior to 1927, the family members, including descendants, should be beyond the jurisdiction of the current Act as they are exempted by Article 5 of Chapter II of the Act (GoB 2000). Therefore, this study treated families claiming residence in this area for an indefinite period, or more than 100 years, as legal dwellers, even though they could not document their claims.

Indigenous people constitute the third category of legal residents. This general study area comprises the traditional territory of the Chakmas (Karim 2008; Rahman 2011). There are several Chakma settlements in this area, with occasional movement between them. Therefore, this study included all Chakma households as legal residents of the reserved forest area, regardless of their status as forest villagers and the duration of their residence.

To sum up, whereas all Chakma households were treated as legitimate forest dwellers, all Rohingya households were considered encroachers, because they are very recent refugees. Of the 139 Bengali households (Table 8.2), 60 households either belonged to the category of forest villagers or had settled in this area a long time ago before the Forest Act was implemented and were therefore legitimate residents. However, the remaining 79 Bengali households had no legal grounds for living inside the TWS and were therefore treated as encroachers. There were a total of 86 legitimate households and 85 encroaching households within the three settlements located inside of the TWS.

8.3.3 Household Incomes

Six categories of household income were separately recorded. These categories were farming, betel leaf (*pan*) cultivation, business, daily wage labor, fishing, and other miscellaneous activities. As described in previous chapters, because betel leaf cultivation (hereafter called *pan*) is a major means of subsistence in Cox's Bazar District and because it negatively affects the state of forests through the consumption of forest products required to build the *pan boroj* cultivation facilities (Nishorgo Support Project 2006; Rahman et al. 2013; Tani et al. 2011; Tsuruta et al. 2012), income from pan cultivation is considered separately from other categories of farming income in this analysis. For the farming and *pan* cultivation categories, the respondents were asked to estimate their gross sales of products and the costs of engaging in these activities. Their incomes were calculated by subtracting these costs from sales. Figures for incomes calculated for each category were summed to obtain total annual household incomes (see Table 8.3). The mean total annual

Table 8.3 Mean annual income sources (in thousands of Bangladesh Taka (BDT)) of the dweller groups

	Farming	*Pan*	Business	Labor	Fishing	Other	Total
Chakma	8.8	19.3	0.5	21.3	0.0	0.8	50.7
Bengali (L)[a]	38.8	46.7	29.4	30.6	10.6	23.7	179.7
Bengali (E)[a]	14.5	30.0	20.9	17.7	0.0	25.5	108.6
Rohingya	7.9	0.0	4.4	68.4	0.0	3.3	84.0
Insiders total	21.9	33.2	20.2	24.6	3.7	20.3	123.9
Outsiders	11.4	9.5	46.2	30.3	42.5	23.8	163.7
All	18.4	25.3	28.8	26.5	16.6	21.5	137.1

[a]Bengali (L), legal dwellers; Bengali (E), encroachers

income of all households residing within the TWS was 123,900 BDT. Of the "insider" groups, legitimate Bengalis were found to be economically much better off than other groups of insiders. Chakma households earned the lowest incomes at less than half of the mean total household income (Table 8.3). Notably, Rohingya households, despite their encroacher status, earned 60% more than Chakma households. The mean household income of the Bengali encroachers was 40% less than that of the legitimate Bengali forest-dwelling households. The difference between the mean household incomes of the Bengali encroachers and legitimate Bengalis was statistically significant, as assessed through a *t*-test at a 5% level of significance ($t = 3.4383$, $p = 0.0008$). The mean income of residents living outside the TWS was higher than that of residents, in general, living inside the TWS, but it was comparable to the mean household income of legitimate forest-dwelling Bengalis. The difference between mean household incomes of legitimate Bengalis and outsiders was not statistically significant, as assessed with a *t*-test at a 5% level of significance ($t = 0.4315$, $p = 0.6668$).

Table 8.4 shows the proportions of income earned from different activities. Chakma households, whose incomes were lowest, were heavily dependent on *pan* cultivation and daily wage labor, which each accounted for approximately 40% of their incomes. The incomes of legitimate Bengali households earned from different activities were similar, with the main emphasis being on *pan* cultivation, rice farming, and other types of farming. Bengali households belonging to the encroacher category earned less than legitimate Bengali households, but the proportions of incomes from different sources were similar among both groups. Daily wage labor accounted for 81% of incomes earned by Rohingya households. This indicates that Rohingya households lack resources such as land and property, which can reasonably be attributed to the fact that they are recently arrived refugees.

There were no clear differences in income sources among Chakma, legitimate Bengali, Bengali encroacher, and Rohingya households. Regardless of their legal status, households belonging to minority ethnic groups tended to depend on daily wage labor, presumably because they lacked other resources such as land.

A clear contrast between the subsistence patterns of the insider and outsider households was apparent. Dependence on farming, including *pan* cultivation, was

Table 8.4 Proportions of total household income resulting from different activities

	Farming (%)	*Pan* (%)	Business (%)	Labor (%)	Fishing (%)	Other (%)
Chakma	17	38	1	42	0	2
Bengali (L)[a]	22	26	16	17	6	13
Bengali (E)[a]	13	28	19	16	0	24
Rohingya	9	0	5	81	0	4
Insiders total	18	27	16	20	3	16
Outsiders	7	6	28	19	26	15
All	13	18	21	19	12	16

[a]Bengali (L), legal dwellers; Bengali (E), encroachers

Table 8.5 Number of households by group and length of habitation (years)

	>10	10–24	25–49	50–99	≤100	Total[a]
Chakma	2	2	3	9	9	25
Bengali (L)[b]	7	8	13	12	18	58
Subtotal (L)[b]	9	10	16	21	27	83
Bengali (E)[b]	14	27	23	10	0	74
Rohingya	6	1	0	0	0	7
Subtotal (E)[b]	20	28	23	10	0	81
Outsiders	5	7	14	12	41	79

[a]This table only includes households whose duration of residence is on record. Therefore, seven households located within the TWS and six households located outside the TWS were excluded from this analysis

[b]Bengali (L), legal dwellers; Bengali (E), encroachers; subtotal (L), legal dwellers; subtotal (E), encroachers

much greater among households located within the TWS (45%) than among those located outside (13%). By contrast, fishing was much more economically important for households living outside the TWS, presumably because Kader *para* is located near the coast. Incomes obtained from business activities were also greater for those households located outside the TWS compared with those located inside the sanctuary.

8.3.4 Residential History

Legal dwellers have evidently stayed longer in the TWS than illegal dwellers, with more than 50% of the Chakma and legitimate Bengali households having lived in the TWS for a period exceeding 50 years. By contrast, encroacher households (Bengali encroachers and Rohingya) appeared to be relatively new in the area (see Table 8.5). More than 50% of Bengali households belonging to the encroacher category, and all Rohingya households, have been resident in the area for periods of less than 25 years. The duration of residence of those households living outside the TWS was similar to that of legal residents within the TWS, but many households living outside the TWS demonstrated longer residential histories than those located inside the TWS.

8.4 Discussion

Based on the above analysis, the following summary profiles of local groups of residents within and outside of the TWS are presented. About half of the forest dwellers in the reserved area are encroachers. These encroachers are mostly Bengalis, but some are also Rohingya refugees from Myanmar. Legitimate dwellers include both Chakma and Bengali households. Among the forest dwellers within the TWS, legitimate Bengali households earn the highest incomes, while illegitimate Bengali households earn significantly less than their legal counterparts. The incomes of households belonging to the two minority groups, the Chakma and Rohingya, are even lower than those of Bengali encroachers. Whereas the overall average income of households located inside the TWS was lower than that of households located outside the sanctuary, the mean income of legitimate Bengali households was comparable to that of the outsiders. The subsistence activities of households within and outside the TWS differed. Households located inside the TWS were more intensively engaged in agriculture, whereas those located outside were more intensively engaged in fishing and business for income generation. Households belonging to minority groups within the TWS relied heavily on daily wage labor for their living.

There are several implications of these findings relating to deforestation on the Teknaf Peninsula. First, a large number of people live inside the TWS. These people affect forests by consuming forest products for various purposes such as fuelwood and building materials. Forest dwellers also clear portions of the forests to build homesteads. Moreover, their major means of subsistence is farming, which has a greater impact on the state of the forest than do other occupations such as small businesses and fishing, which are more widely practiced by residents outside the TWS. Households engaged in farming create agricultural fields by clearing forested areas. Moreover, *pan* cultivation consumes further forest products that are needed to construct *pan boroj* (see Chaps. 3, 4, 5, and 6 of this volume; Tani et al. 2011; Tsuruta et al. 2012).

A second important finding is that at least half of the people living inside the TWS are low-income encroachers. Although Rohingya refugees are sometimes blamed for forest degradation in Teknaf (Rahman et al. 2013; Uddin and Khan 2007), there are just a few Rohingya households located in the study area, whereas there are many households belonging to the Bengali encroacher category in this area. Moreover, there continues to be a steady flow of new encroachers into the area. Twenty-five percent of all encroachers have lived in the TWS for less than 10 years, and more than 50% of them have lived there for less than 25 years. The household incomes of these encroachers are 40% lower than those of legitimate Bengali households. This may reflect the fact that the poorer segments of society struggle to survive, attempting to take advantage of “free” resources available in the reserved forest to make a living. In other words, poverty is a major driving force of deforestation in this area. Therefore, if more people become impoverished, some of them may have no choice but to move into the forest to use freely available resources.

8.5 Conclusions

This chapter has shown that a sizable population resides in the restricted forest area. This population factor appears to be a major driving force of forest degradation. The analysis of the village case study revealed the characteristics of its inhabitants who live inside the boundaries of the reserved forest area in the TWS. It showed that half of the residents living in the reserved forest area are not legally entitled to live there. However, the Bangladesh Government is seemingly unable to effectively control people’s movements into reserved areas, as evidenced by the steady and continual inflow of encroachers into the area. There is a growing concern that new settlers in the reserved areas will contribute to further degradation of already damaged forests because of the very presence of their homesteads and the environmental pressure resulting from the cultivation of *pan* and other crops in the forest. Thus, it is crucial to control these human activities to enable the forests to regenerate in the near future.

Because this case study only examined a small section of the area under investigation, it would be necessary to expand the characterization of encroachers in a future study. Whereas a tentative conclusion of the case study analysis is that poverty is a contributing factor to forest encroachment, it is also necessary to examine forest conditions on the eastern side of the peninsula and/or in areas south of the TWS where the magnitude of forest encroachment is much lower.

References

Akhter S, Sohel MSI, Rana MP, Alamgir M (2009) Impact of forest and non-forest villagers on Ukhia and Inani forest range under Cox’s Bazar (south) forest division. Bangladesh Proc Pakistan Acad Sci 46(1):13–22

Gibson CC, McKean MA, Ostrom E (2000) Explaining deforestation: the role of local institutions. In: Gibson CC, McKean MA, Ostrom E (eds) People and forests. MIT Press, Cambridge, pp 1–26

GoB (2000) Government of Bangladesh. Forest Act, 1927 (updated 30 April, 2000)

IPAC (2009) Site-level field appraisal for protected area co-management: Teknaf Game Reserve Cox’s Bazar. IPAC-Southeastern Cluster

IPAC (2011) Land use change trend analysis in seven protected areas in Bangladesh under IPAC through application of Landsat images. USAID

Karim MN (2008) Local knowledge of indicator birds: implications for community-based ecological monitoring in Teknaf game reserve. In: Fox J et al (eds) Connecting communities and conservation: collaborative management of protected areas in Bangladesh. East-West Center, Honolulu, pp 139–160

Nishorgo Support Project (2006) Management plans for Teknaf Game Reserve. Nishorgo Support Project

Rahman MA (2011) Deforestation and forest conservation in a Tanchangya community. In: Fox J et al (eds) Rural livelihoods and protected landscapes: co-management in the wetlands and forests of Bangladesh. East-West Center, Honolulu, pp 118–132

Rahman MZ, Tani M, Moslehuddin AZM, Ullah SMA (2013) Use of information sources in maintaining livelihoods by Rohingya refugees around Teknaf Wildlife sanctuary. J Agrofor Environ 7(2):15–18

Tani M, Rahman MZ, Rahman MA, Asahiro K, Akhter S (2011) Deforestation by daily activities in the Teknaf Peninsula, Bangladesh. Proceedings of the International Conference on Environmental Aspects in Bangladesh (ICEAB) 2011, pp 73–75

Tsuruta H, Tani M, Rahman MZ (2012) The study on the impact of betel leaf cultivation on the local forest in the Teknaf Peninsula. Proceeding of 3rd International Conference on Environmental Aspects of Bangladesh, pp 51–53

Uddin MS, Khan MASA (2007) Comparing the impacts of local people and Rohingya refugees on Teknaf game reserve, Bangladesh. In: Fox J, Bushley B, Dutt S, Quazi SA (eds) Making conservation work: linking rural livelihoods & protected area management in Bangladesh. East-West Center and Nishorgo Program of the Bangladesh Forest Department, Dhaka, pp 149–175

Chapter 9
Livelihoods of Rohingyas and Their Impacts on Deforestation

M.Z. Rahman

Abstract This chapter focused primarily to reveal a comprehensive picture of the livelihood mechanisms of undocumented Rohingyas and their possible impacts on the forest and other natural resources, including the Teknaf Wildlife Sanctuary (TWS). A cluster of Rohingyas with this status were selected for the study. This group lives along the Bay of Bengal coastline in the village of Shamlapur in Teknaf Upazila, which falls within the district of Cox's Bazar. A sample of 125 household heads was randomly selected from a total of 980 households. Additionally, data were collected from 65 Rohingya household heads of Kerontoli village of Teknaf, which is located alongside the Naf River and the TWS. The Rohingya of this area lives among the local community. Data were primarily generated from interviews and focus group discussions. The livelihood status of most of the Rohingyas was found to be very low, with their occupations resulting in the destruction of natural resources. Status of livelihoods regarding all the capitals was very low for Rohingyas of both the study sites. Illegal migration of Rohingya to Bangladesh and their immediate job opportunities in the vicinity of deweling places dertmined their livelihood patterns. Random fishing in water bodies (sea and river) and cutting woods from hills exerted enormous negative impacts on sea and TWS. Moreover, the exhaustive livelihoods created bad competitions with local dwellers. The analysis indicated that issues of concern for the Rohingyas could only be mitigated or effectively resolved through cordial talks held between Bangladesh and Myanmar on these issues.

Keywords Rohingya • Forest resources • Teknaf Wildlife Sanctuary • Livelihood • Refugee • Migration

M.Z. Rahman (✉)
Department of Agricultural Extension, Faculty of Agriculture, Bangladesh Agricultural University, Mymensingh, Bangladesh
e-mail: zulfikarbau64@gmail.com

M. Tani, M.A. Rahman (eds.), *Deforestation in the Teknaf Peninsula of Bangladesh*, DOI 10.1007/978-981-10-5475-4_9

9.1 Introduction

The Rohingyas are an ethnic, religious, and linguistic minority in Myanmar who primarily inhabit the three townships of North Arakan, which borders Bangladesh. Their number in North Arakan is estimated to be approximately 725,000 (Lewa 2010). Professing Islam and of South Asian descent, they are related to the Chittagonian Bengali that live across the border but are distinct from the majority population of Myanmar (formerly Burma), who are of Southeast Asian origin and mostly Buddhist. They gradually faced exclusion in Myanmar, especially since the military takeover in 1962. They were rendered stateless by the Burma Citizenship Law of 1982, which mainly confers the right to citizenship on members of the 135 "national races" listed by the government and which does not include the Rohingyas. Denial of citizenship is the key mechanism underlying the exclusion of institutional discrimination against this group. The Rohingyas face severe restrictions on their movement and marriages, arbitrary arrests, extortion, forced labor, and confiscation of land. The combination of restrictions and abuses has had a dramatic impact on their economic survival (Mollah et al. 2004).

Hundreds of thousands of stateless Rohingyas have fled brutal oppression in Myanmar. Impoverished Bangladesh witnessed two mass exoduses of 250,000 Rohingya refugees in 1978 and 1991/1992, which were followed by forced repatriation. About 28,000 *Rohingyas* remain in two precarious refugee camps assisted by the United Nations High Commissioner for Refugees and a few nongovernmental organizations (NGOs) (Sajjad 2003). However, the exodus of Rohingyas never stopped, and new arrivals do not have access to the existing refugee camps, and there is no mechanism for them to seek protection.

An estimated 200,000 unregistered Rohingya refugees have settled among the local population in Teknaf, mostly in slums and villages throughout Cox's Bazar District, but also in smaller numbers in the Chittagong Hill Tracts, where they eke out a hand-to-mouth existence without any humanitarian assistance and are vulnerable to exploitation and arrest. While keeping them invisible and portraying them as economic migrants, Bangladesh has generally tolerated their presence, but anti-Rohingya sentiments have steadily grown among the local population as a result of manipulation by the local political elite and the media (Anonymous 2015).

In 2002, in response to threats of eviction, an initial makeshift camp was established near Teknaf, and the 10,000 residents were ultimately relocated to a new site in Leda in June 2008. In early 2008, a new makeshift camp started sprawling around the Kutupalong refugee camp as a consequence of eviction threats against self-settled Rohingyas during the voter registration that preceded the national election of December 2008. As the population increased, sections of the makeshift camp were demolished by the Bangladesh authorities on three occasions in June and July 2009 (Bari and Dutta 2004).

In parallel, at the end of 2007, Bangladesh law enforcement agencies started arresting and repatriating Rohingyas across the border to Myanmar. Initially, only new arrivals were targeted, but since mid-2009, self-settled refugees have also been deported. To avert deportation in recent days, Rohingyas have been trying to settle in areas where social and physical camouflages can be achieved easily. They mix socially with Bangladeshi communities by renting their houses and living there. Sometimes they enter into matrimonial relationships with the local people to settle permanently. However, they also live in huts in remote, government-owned jungles (Bari and Dutta 2004; Mollah et al. 2004). In this study, both Rohingya populations, those living in the remote forests and local communities, have been considered. It is assumed that their different locations and strategies of living in Bangladesh will reveal a comprehensive picture of their livelihood mechanisms and their possible impacts on the forest and other natural resources, including the Teknaf Wildlife Sanctuary (TWS) (GoB 2009).

9.2 Methodology

The study was conducted in Shamlapur village of Teknaf Upazila of Cox's Bazar District. A cluster of Rohingya people live in the buffer forest of the coastal belt of Shamlapur. This village is a local marketplace that plays an important role in the locality because it is far away from the main markets of Teknaf and Cox's Bazar. Moreover, the TWS, previously known as the Teknaf Game Reserve, splits the community longitudinally. The local people stated that the total number of Rohingya households and population are about 980 and 5880, respectively, in the buffer forest site in Shamlapur. One hundred and twenty-five (approximately 13%) randomly selected Rohingya heads of household were interviewed using a structured interview schedule to assess their livelihood status. Secondary data regarding different community- and service-oriented information were collected from different sources. Data were collected during March 2012. Focus group discussions (FGDs) were also conducted to assess their social and natural capital, as well as service delivery issues, which affect their livelihoods. Again, two FGDs were conducted to explore the correlates of their livelihood status. The outcomes of the two FGDs were merged to achieve a comprehensive result.

Additionally, data were collected from the Kerontoli village of Teknaf, which is located alongside the Naf River and the TWS. The Rohingya of this area lives among the local community. There were approximately125 Rohingya families here, and data were collected from 65 randomly sampled Rohingya heads of household, as was done for the other site. The findings of the two study sites were analyzed, merged in some cases, and compared to achieve a comprehensive picture of the livelihood issues of Rohingyas living around the TWS, as well as their impacts on natural resources, especially on forests.

Livelihood status was measured based on the extent of their livelihood assets. Assets referred to the resource base of the people. Assets are often represented as a pentagon in the sustainable livelihood framework, consisting of the following five categories: natural resources (also called natural capital), physical goods (physical capital), monetary resources (financial capital), manpower with different skills (human capital), and social networks of various kinds (social capital) (FAO 2009).

These categories are related to the following types of issues addressed by this study. Human capital represents the skills, knowledge, ability to work, and good health that together enable people to pursue different livelihood strategies and achieve their livelihood objectives. It is an aggregate economic view of humans acting within economies, which is an attempt to capture the social, biological, cultural, and psychological complexity as they interact in social and/or economic transactions. Some service delivery indicators of human capital that were considered in this study are the public expenditure per capita, the number of physicians per thousand population, and the primary school student to teacher ratio. The outcome indicators considered are life expectancy at birth, the under-five mortality rate, the primary school enrolment rate, the school completion rate, and the literacy rate.

9.3 Results and Discussion

9.3.1 Reasons for Migration from Myanmar to Bangladesh

The story of the Rohingya begins with their exodus to Bangladesh. Under certain unusual conditions, the Rohingyas have been immigrating to Bangladesh over the last few decades. Until approximately one decade ago, the matter of the migration of Rohingyas to Bangladesh had been considered somewhat leniently by the government of Bangladesh; however, the government has adopted a stricter policy toward migration recently. Whatever their standing in Bangladesh, Rohingyas have reported several major reasons for their migration to Bangladesh. These were:

- Torture of the majority ethnic groups, such as the Monand and others by the *Nasaka*, the border force in Myanmar
- Putting Rohingyas in jails and subjecting them to forced labor
- Insecurity of life and assets in Myanmar
- Religious dissimilarity and discomforts in Myanmar
- Migration of kin and clans to Bangladesh
- Cultural similarity and political peace in Bangladesh

During the discussions, the Rohingya respondents reported that their main reasons for leaving Myanmar were the merciless torture by the *Nasaka* and local ethnic groups. Torture, as well as killing, rape, jailing, forced labor, and burning houses and fields, was reported to be common in Myanmar. Thus, they felt that their lives and properties were insecure in Myanmar, which led them to flee to Bangladesh.

9.3.2 Key Sociodemographic Features of Rohingya Heads of Household

As settlers, the overall socioeconomic standards of the Rohingyas were greatly inferior to those of the local people. Among others, their lack of education was poor; on average, the Rohingya at the buffer forest site only had 2 years of education, while

Table 9.1 Selected sociodemographic features of the Rohingyas

Characteristics (with means of measurement)	Observed range		Mean	
	Buffer forest	Kerontoli	Buffer forest	Kerontoli
Age (years)	22–70	18–80	42	36
Education (years of schooling)	0–10	0–12	2	1
Household size (no. of members)	3–10	1–11	6	5
No. of laborers/household	1–4	1–3	1.3	1.5
% labor force of the household	16.7–66	16–57	20.8	16.8
Annual household income (000s of Bangladeshi Taka (BDT)	29.2–109.5	18–873.6	33.6	101.9
Time living in Bangladesh (years)	2–13	0.2–40	3.6	13.3
Length of fishing experience (years)	1–11	0–18	4.7	0.5
Social mobility (rated score, possible score 0–24)	0–8	0–12	5	8

those at the Kerontoli site had 0.95 years (Table 9.1). It is remarkable that this length of schooling predominantly took place in religious schools. Only a few heads of household were educated in private, preprimary schools, as they were denied access to any kind of government services and facilities. Some of the Rohingyas obtained Bangladesh national identification cards through improper channels. This allowed them to enroll in public schools, and it opened avenues to using other public facilities in Bangladesh. This means that they are treated as legal Bangladeshi citizens.

The number of earning members per family was low, as was their annual household income, although the income was slightly higher at the Kerontoli site than the buffer forest site, as the adult male members at the former site had an opportunity to work at a nearby river port. However, the Rohingyas living in the buffer forest generally had only one livelihood option, i.e., fishing in the sea.

9.3.3 *Overall Scenario of the Livelihood Status of Rohingyas*

The service delivery indicators for both areas indicate that the Rohingyas living in Bangladesh are in a helpless condition in which public support and services are absent. The following issues were common to both sites:

- Per capita public expenditure: 0
- Physicians per thousand population: 0
- Government primary school student to teacher ratio: 0

The outcome indicators that differed between the two sites are noted below (figures in parenthesis are for the Rohingyas of Kerontoli):

- Life expectancy at birth: 60.4 years (66.2 years)
- Under-five mortality rate: 6.5% (5.8%)
- Religious school (class five level) enrolment and completion rate: 2% (primary school, 5%); completion rate 1% (2%)

- Literacy rate: 40% (45.5%)
- Newspaper readership: 2% (5.2%)

Thus, it was observed that the overall livelihood indicators for the community were very poor and undesirable. As the Rohingyas are denied any access to public facilities, their livelihood levels were likely to be very poor. Their stateless situation may lead them to further engage in low-paying livelihood activities. In the developing world, exposure to the greater economic community is important for achieving livelihood motivation and its implementation. However, the Rohingyas were compelled to remain in their local communities, as they were illegal residents. Other studies (Karim 2005; Rahman et al. 2013) also support these findings.

9.3.4 Status of Rohingyas Based on Five Types of Capital

Livelihood status was calculated based on five types of capital. Possible livelihood status scores ranged from 28 to 84, and the observed scores ranged from 28 to 34, while the average score was 31.2. As shown in Table 9.2, it was observed that all of the Rohingyas had a low livelihood status. This manifestly expresses their stateless situation in their present living areas. Finding no alternative to their existing situations, they were compelled to embrace their present livelihood status. The small differences in the livelihood status between the two sites were not significant.

Labor status, health and nutritional status, skills, and knowledge were considered to be core indicators of human capital. Each of the indicators was rated by the respondents as low, moderate, or high, with respective scores of 1, 2, and 3; possible scores for all of the indicators ranged from 4 to12. For the Rohingya living in the buffer forest, the scores ranged from 4 to 6, with a mean of 4.2 (Table 9.3), while the scores for the Rohingya living in Kerontoli ranged from 4 to 6, with a mean of 4.5. These results indicate the low livelihood status of the Rohingya at both sites.

Natural capital is the term used for the natural resource stocks from which resource flows and services that are useful for livelihoods are derived. Access to land, water, wildlife, marine resources, forests, and storm protection were mainly considered to be measures of natural capital. Each of the indicators was rated by the respondents as low, moderate, or high, with respective scores of 1, 2, and 3; possible cumulative scores ranged from 6 to 18.For Rohingyas living in the buffer forest, the

Table 9.2 Overall score for livelihood status of Rohingyas (n = 190)

Status of livelihood (score)	Rohingya						Mean score value		
	Buffer forest		Kerontoli		Overall		Buffer forest	Kerontoli	Overall
	No.	%	No.	%	No.	%			
Low (28–47)	125	100	65	100	190	100	32.1	30.3	31.2
Medium (48–65)	0	0	0	0	0	0	0	0	0
High (66–84)	0	0	0	0	0	0	0	0	0

Table 9.3 Status of various types of capital used to determine the livelihood scores

		Observed score		Mean value	
Types of capital	Possible score	Buffer forest	Kerontoli	Buffer forest	Kerontoli
Human	4–12	4–6	4–6.5	4.2	4.5
Natural	6–18	6–7	6–8.1	6.3	6.5
Social	6–18	6–7	7–8	6.3	6.9
Physical	7–21	7–8	7–8.2	7.6	8
Financial	5–15	5–6	5–6.8	5.5	6

natural capital scores ranged from 6 to 7, with a mean of 6.3. For the Kerontoli Rohingyas, the natural capital scores ranged from 6 to 8.1, with an average of 6.5. Like the other types of capital, the level of natural capital was very low for both sites.

Social capital is the expected, collective, or economic benefits derived from preferential treatment and cooperation between individuals and groups. It refers to those stocks of social trust, norms, and networks that people can draw upon to solve common problems. It is mediated through kin networks and group membership. In other words, social capital is taken to mean the social resources upon which people draw in pursuit of their livelihood objectives. There is no single measure of social capital. Social capital includes dimensions such as trust in government, memberships in civic organizations, hours spent volunteering, voter turnout, newspaper readership, and access to basic services. Each of the indicators was rated by the respondents as low, moderate, or high, with respective scores of 1, 2, and 3; the possible scores ranged from 6 to 18. The observed scores for Rohingyas living in the buffer forest ranged from 6 to 7, with a mean of 6.30 (Table 9.3). The observed scores for the Kerontoli Rohingyas ranged from 7 to 8, with a mean of 7.6. Although the Kerontoli Rohingyas were slightly higher than those living in the buffer forest, the level of social capital was low for both sites.

Physical capital comprises the basic infrastructure and producer goods needed to support livelihoods. Houses, vehicles, equipment, livestock, access to information, an adequate water supply, and sanitation were considered to be measures of physical capital. Each of the indicators was rated by the respondents as low, moderate, or high, with respective scores of 1, 2, and 3; the possible scores ranged from 7 to 21. The observed human capital scores of the Rohingyas living in the buffer forest ranged from 7.0 to 8.0, with a mean of 7.6. The scores for the Kerontoli Rohingya inhabitants ranged from 7.0 to 8.2, with a mean of 8.0.

Financial capital denotes the financial resources that people use to achieve their livelihood objectives. Savings, gold/jewelry, access to regular income, and access to credit and insurance were considered to be indicators of financial capital in this study. Each of the indicators was rated by the respondents as low, moderate, or high, with respective scores of 1, 2, and 3; possible scores ranged from 5 to 15. The scores of the Rohingyas living in the buffer forest ranged from 5 to 6, with a mean of 5.5. Scores of the Kerontoli Rohingyas ranged from 5 to 6.8, with a mean of 6, which indicated that they had an economically disadvantaged status.

Increasingly, it is recognized that in addition to these five categories, it is important to include an analysis of political capital. This goes beyond social capital, in that an individual's stock of political capital will determine his/her ability to influence policy and the processes of government. An understanding of political capital is important in determining the ability of households and individuals to claim rights to assistance. The study observed that Rohingyas had been passing their lives in stranded and stateless conditions in which they did not possess any ability to influence any aspect of the society or its politics. In the same way, they could not claim any right to receive assistance from government facilities.

9.3.5 Correlates of Rohingyas' Livelihoods

Correlates of the livelihoods of Rohingya refugees were explored qualitatively based on the FGDs, in which eight people were involved in each of the group discussions. Separate discussions were held for each site, and a checklist of relevant matters was used to conduct the FGDs. Rohingyas maintain a very indigent life with limited or no options for livelihood maintenance. In this investigation, although qualitatively measured, issues affecting their helpless livelihood condition were revealed (Table 9.4).

Table 9.4 Issues affecting the livelihood of the Rohingyas

	Study areas			
	Buffer forest		Kerontoli	
Issues associated with livelihood	*Impact trend*	*Rank order*	*Impact trend*	*Rank order*
Illegal migration of Rohingya to Bangladesh	–	1	–	4
Nonrecognition as refugee by the Bangladesh Government	–	2	–	5
Acceptance of Rohingyas by the local people	+	3	–	3
No need for capital or other materials to fish in the sea	+	4	+	13
Easy access to coastal resources (mainly fish and shrimp fry)	+	5	+	12
Access to nearby hills, although this is illegal	+	6	+	2
Lack of access to the primary education system	–	7	–	11
Religious compatibility in Bangladesh	+	8	+	7
Acute shortage of options for labor utilization	–	9	–	14
Similar dialect/language in the locality	+	10	+	6
No access to any government institution	–	11	–	8
No support from the Bangladesh Government	–	12	–	15
Harassment by the Bangladesh police	–	13	–	16
Restricted movement in the locality for earning	–	14	–	9
No chance to return to Myanmar	–	15	–	10
Ability to work at the nearby river port	+	16	+	1

Illegal migration of Rohingyas to Bangladesh and nonrecognition as refugees by the Bangladesh Government were the two major deleterious issues affecting the livelihoods of the Rohingya living in the buffer forest. While well accepted by the local people, the lack of a need for capital or other materials to fish in the sea were the two major issues that positively affected their livelihoods. On the Kerontoli site, the factors that negatively affected their livelihoods were illegal migration to Bangladesh and the acceptance of the Rohingyas by the local people. Ability to work at the nearby river port and access to a nearby mountain, although illegal, were the two main issues that positively affected their livelihoods. Acceptance by the local people and the ability to earn income were important factors that affected the livelihoods of the Rohingyas at both sites.

It was revealed that there were differences in the order in which the issues were ranked in the two locations. While the "illegal migration to Bangladesh" factor was ranked first in the buffer forest area, it was ranked fourth in Kerontoli. In contrast, "ability to work at the nearby river port" was ranked first at the Kerontoli site, while it was ranked last at the buffer forest. The differences in the rankings result from the fact that the issues concerning the livelihoods of illegal Rohingyas are related to the nature of their work and their security in the areas where they live.

9.3.6 Livelihood Options Adopted by Rohingyas

As already identified, there remain various factors that affect the livelihoods of the Rohingyas, which led them to unsuitable livelihoods. These factors directed them to adopt one or more of the following livelihood options (Table 9.5). Rohingyas living in Kerontoli were mostly port laborers. This is the reason that they did not or could not go to an official refugee camp and, thus, sheltered themselves in local communities near the port. Additionally, their living community is very close to the TWS, and some of them have been living in the TWS that is owned by the Bangladesh Forest Department (Khan et al. 2009). Other than working mainly at the port, some of them had been cutting trees from the hills to sell in the market, as well as fishing in the river.

Table 9.5 Livelihood activities adopted by the Kerontoli Rohingyas

	Rohingyas	
Livelihood options adopted	Percentage	Ranking
Port labor	45.4	1
Cutting fuelwood from hills and selling it in the market	23.8	2
Fishing in the Naf River	12.3	3
Working as a translator in the port	9.2	4
Small business/hawking goods	6.2	5
Working abroad	1.5	6
Day labor (nonfarm activities)	1.5	7

Table 9.6 Livelihood activities adopted by the Rohingyas in the buffer forest

	Rohingyas	
Livelihood options adopted	Percentage	Ranking
Fishing in the sea	36.8	1
Cutting fuelwood from hills	16.0	2
Cultivating sun grass on hills	14.4	3
Collecting small trees from hills	12.8	4
Day labor	08.8	5
Small grocery/hawking goods	04.0	6
Fishing boat renting	02.4	7

Additionally, approximately two-thirds of them (either the family head or any other member of the household) used to cut trees and twigs in TWS for use as domestic fuel. This is why they have been linked directly to deleterious effects on natural resources and the environment (Chattarjee 2011; Rahman et al. 2011; Uddin and Khan 2007).

Table 9.6 reveals that approximately one-third of the Rohingya family heads were engaged in fishing in the sea. Collectively, 43% of them were directly involved in activities that destroy the hill forests. Various studies found similar results regarding forest destruction by the Rohingyas in and around the Teknaf hill forest (Chatterjee 2011; Lewa 2010; Tani et al. 2011). An interpretation of the findings demonstrates that approximately 80% of the heads of household were engaged in resource-depleting activities in the sea or hills.

9.3.7 Livelihood Activities of Rohingyas and Their Impacts on Deforestation

Rohingyas adopted environmentally degrading livelihood activities because of their lack of economic opportunities. They realized that their involvement in these activities led to adverse effects on natural resources and the environment. Some of them live in the buffer forest and some in a government-owned sanctuary (i.e., the hill forest). Their illegal homesteads and livelihood activities damaged the forest (Table 9.6). Construction of homes in the hill forests by encroachers transforms the natural forest ecosystem to a man-made ecosystem. Specifically, the Rohingyas living in the buffer forests along the seaside directly damage the trees planted by the government. Studies by Mukul et al. (2008) and IPAC (2009) also showed similar outcomes in their investigations. To meet their basic household need for fuel, the Rohingyas depend solely on biomass fuel sources, which they obtain by tree felling and twig collecting in the TWS. Approximately 87% of the Rohingyas in the buffer forest either cut trees, twigs, and branches of trees or cultivate sun grass on the hills of the TWS (Table 9.7). It is worth mentioning that the Rohingyas also purchase fuelwood from the market or neighbors.

A study of the hill deforestation by Rahman et al. (2011) showed that "maintaining pan boroj" and "cutting trees from the hill" were the first and second most,

Table 9.7 Consequences of the Rohingyas' livelihood activities

Livelihood activities	Consequences observed
Living in a government-owned buffer forest and the TWS	Destruction of the forest, pollution of the environment, government oppression
Encroachment in government-owned forests	Destruction of trees and other wildlife, habitat destruction, land erosion, soil deposition in the streams and sea
Year-round collection of fish from the river and sea	Extinction of aquatic resources, reduced catches, risk of injury or death, decrease in income

respectively, environmentally consequential activities of the local people. This is due to the inhabitants' lack of knowledge of the environment. People are making pan boroj to cultivate betel leaf to obtain financial benefits, which changes the ecology of the area. Because pan boroj is made by clearing forests and other vegetation on the hill, their construction destroys the forests and vegetation and seriously degrades the hill environment.

Approximately one-fourth (24%) of the Rohingyas at the Kerontoli site, finding no other livelihood options, cut twigs, trees, and their branches and sell them in the markets to meet their household expenditures. All of these actions cause severe deforestation in the TWS, which leads to habitat loss and severe landslides and erosion, which increase pollution. Landslides from the hill area are major cause of soil deposition on the fertile crop fields of the TWS, and they result in sediment runoff into the streams and sea.

Cutting and selling wood by the Rohingya, as a livelihood activity, should not be viewed as the only cause of deforestation, as 86% of the Kerontoli Rohingyas also cut trees, their branches, twigs, and leaves, as well as bushes, from the hills for household use as biomass fuel. The percentage was slightly lower (56%) in the case of the buffer forest Rohingya. Thus, it is recognized that the deforestation caused by the Rohingya resulted from their need to maintain their livelihood activities and to obtain a supply of biomass fuel for their households.

The year-round indiscriminate catching of fish from the river and sea depletes aquatic resources, including diverse species of fish. This leads to reduced catches and a decrease in income from fishing. Moreover, the conventional methods used by the fishermen result in the risk of injury or death. Accordingly, the environmentally degrading livelihood activities of the Rohingyas pose a threat to natural resources such as forests and aquatic systems, although the total environmental effect of the Rohingya is much less than that of the local inhabitants. A number of mitigation measures to resolve these matters have been sought by the government and NGOs. Government policies were implemented to generate alternative sources of income for the Rohingyas. Alternative income-generating activities include cattle fattening, growing vegetables on rented land, small trade, poultry rearing, and other small and affordable activities.

Despite their hesitations, the Rohingyas' illicit felling of forest trees and destruction of other natural resources had tremendous adverse effects on natural resource management in the locality, including the TWS. As a result, the Bangladesh Government and other concerned groups should consider several issues that might mitigate the destruction of natural resources. These include:

- The Bangladesh Government's agreement to accept Rohingyas as refugees
- The Bangladesh Government's agreement to accept Rohingyas as citizens
- The allocation of special funds and food for the Rohingyas
- Lifting the ban on primary education
- Coverage under the national health plan
- The Myanmar Government's acceptance of the return of the Rohingyas to Myanmar
- Stopping police harassment
- Access to NGO funds
- Establishment of an adult edu-care center in the locality

9.4 Conclusions

Because they were tortured in Myanmar, many Rohingyas took shelter in the southeast districts of Bangladesh, especially in Cox's Bazar District in Teknaf. This began several decades ago. Presently, the refugees are experiencing difficulties, and it is assumed that the matter of the Rohingya refugees might not be resolved immediately, although it has drawn the attention of the international community. This study focused on the livelihoods of the Rohingya and their impact on the deforestation of the TWS. The study revealed some critical concerns regarding their situation, and the following conclusions can be drawn.

Their suffering in Myanmar compelled the Rohingyas to flee to Bangladesh, especially in the southeast districts. In Bangladesh, they have been passing their days in very isolated and wretched conditions. They are completely denied access to social welfare programs, schools, and healthcare, as the Bangladesh Government does not recognize them as refugees. Thus, the local people and the government should allow the Rohingyas to work as laborers in towns and on local farms. The Bangladesh Government, with the help of international agencies, must implement some short-term remedial measures to improve the survival of the Rohingyas.

The low livelihood status of the Rohingyas has been revealed in the study area, although they have been trying their best to live hand to mouth. Their livelihood status could be improved to a fairly satisfactory condition if support could be rendered from the government and allied NGOs.

A very low level of livelihood achievements led the Rohingyas to engage in environmentally degrading livelihood activities, which resulted in deforestation. Along with the local people, they have been destroying the government-owned protected area of the TWS, and the situation has been declining gradually. To alleviate these environmental concerns, the Rohingyas should be made aware of the severe consequences of deforestation and forest degradation. Simultaneously, alternative income-generating initiatives launched by the government and NGOs for the Rohingyas could help alleviate the deforestation problem.

In an overpopulated and resource-poor country like Bangladesh, the Rohingya migration and maintenance of their livelihoods have created tremendous pressure on the government. Natural resource destruction and the decline of law and order in the country are also major concerns. Thus, the problems associated with the

Rohingya could be substantively resolved through bilateral negotiations and the political will of the Myanmar and Bangladesh Governments. International organizations and other relevant communities should attempt to initiate such negotiations.

References

Anonymous (2015) Wikipedia, the free encyclopedia. Rohingya People. Retrieved from: http://en.wikipedia.org/wiki/Rohingya_people. Retrieved on 15 Aug 2015

Bari A, Dutta U (2004) Co-management of tropical forest resources in Bangladesh. Secondary data for pilot protected area: Teknaf game reserve, USAID-Bangladesh and Ministry of Environment and Forest, Govt. of People's Republic of Bangladesh

Chattarjee D (2011) Ecological consequences due to subsistence activities of the people in a hill forest area. MS thesis, Department of Agriculture Extension Education, Bangladesh Agricultural University

FAO (2009) The livelihood assessment tool-kit: analyzing and responding to the impact of disasters on the livelihoods of people. Food and Agriculture Organization of the United Nations (FAO), Rome and International Labour Organization (ILO), Geneva

GoB (2009) Government of Bangladesh gazette notification no. moef/forest-2/02/wildlife/15/2009/492 dt.09/12/2009. Ministry of Environment and Forests, Government of Bangladesh, Dhaka

IPAC (Integrated Protected Area Co-management) (2009) Site-level field appraisal for protected area co-management: Teknaf game reserve, Cox's Bazar. IPAC-Southeastern Cluster, Cox's Bazar

Karim MS (2005) Farmers' use of communication sources in receiving agricultural information. MS thesis, Department of Agriculture Extension Education, Bangladesh Agricultural University, pp 32–33

Khan MASA, Mukul SA, Uddin MA, Kibria MG, Sultana F (2009) The use of medicinal plants in healthcare practices by Rohingya refugees in a degraded forest and conservation area of Bangladesh. Int J Biodivers Sci Mang 5(2):76–82

Lewa C (2010) Unregistered Rohingya refugees in Bangladesh: crackdown, forced displacement and hunger. The Arakan Project, Bangkok. Retrieved from http://oppenheimer.mcgill.ca/Unregistered-Rohingya-refugees-in

Mollah AR, Rahman MM, Rahman MS (2004) Site-level field appraisal for protected area co-management: Teknaf game reserve. Nature Conservation Management (NACOM), Dhaka

Mukul SA, Uddin MB, Uddin MS, Khan MASA, Marzan B (2008) Protected areas of Bangladesh: current status and efficacy for biodiversity conservation. Department of Forestry and Environmental Science, School of Agriculture and Mineral Sciences, Shahjalal University of Science and Technology, Sylhet

Rahman MZ, Tani M, Chattarjee D, Akhter S (2011) Ecological consequences due to subsistence activities of the people in a hill forest area of Bangladesh. Proceedings of ICEAB 2011

Rahman MZ, Tani M, Moslehuddin AZM, Ullah SMA (2013) Use of information sources in maintaining livelihoods by Rohingya refugees around Teknaf Wildlife Sanctuary. J Agrofor Environ 7(2):15–18

Sajjad T (2003) SRI on-site action alert: Rohingya refugees of Burma and UNHCR's repatriation program. Asia Researcher, Survivors' Rights International, Annapolis

Tani M, Rahman MZ, Rahman MA, Asahiro K, Akhter S (2011) Deforestation by daily activities in the Teknaf Peninsula, Bangladesh. Proceedings of International Conference on Environmental Aspects in Bangladesh (ICEAB) 2011:73–75

Uddin MS, Khan MASA. 2007. Comparing the impacts of local people and Rohingya refugees on Teknaf game reserve, Bangladesh. In: Fox, J., Bushley, B., Dutt, S., Quazi, S.A. (eds), Making conservation work: linking rural livelihoods & protected area management in Bangladesh. Dhaka: East-West Center and Nishorgo Program of the Bangladesh Forest Department, pp 149–175

Part III
Scope for Reforestation

Chapter 10
Social Forestry Conditions on the Teknaf Peninsula

K. Asahiro

Abstract This chapter discusses the conditions of social forestry (SF) in the Teknaf and Shilkhali ranges in 2013, 10 years after the SF scheme was first implemented in Teknaf. Field-based surveys and interviews were conducted in September 2013 and in March 2014, and the normalized difference vegetation index (NDVI) was analyzed based on satellite imagery produced in 2010. The correlation matrix was used to describe the relationship between these factors. We found that active tree felling occurred in 70.3% of SFs up to 2013. In eight SFs in which Acacia species had been planted, thinning management deviated from the prescribed SF regulations. Thinning was done properly in 11 SFs, and five SFs with broad-leaved species and seven cleared SFs were identified. These results were mapped, revealing differences in current conditions of the SFs, as well as local measurement requirements. The findings will facilitate SF conservation in the study area.

Keywords Social forest • NDVI • Satellite image • Forest condition • Tree cover

10.1 Introduction

A social forestry (SF) program was started in Bangladesh in 1981 under the community forestry program of the Forest Department (FD). The first formal SF program was initiated in the northwestern districts of Bangladesh in 1981 and 1982 with the assistance of the Asian Development Bank and the United Nations Development Program (Salam and Noguchi 2005). The program was started to protect the forest from encroachment and to encourage local people to participate in conservation activities. After the success of the project, it was disseminated to other regions of Bangladesh.

The need for forest conservation on the Teknaf peninsula has been emphasized in past years because of the importance of forests to biodiversity and local residents'

K. Asahiro (✉)
Department of Environmental Design, Faculty of Design, Kyushu University, Fukuoka, Japan
e-mail: asahiro@design.kyushu-u.ac.jp

M. Tani, M.A. Rahman (eds.), *Deforestation in the Teknaf Peninsula of Bangladesh*, DOI 10.1007/978-981-10-5475-4_10

livelihoods. However, there has been considerable forest degradation in past decades, which prompted the government to create the Teknaf Wildlife Sanctuary (TWS) in 2009. In 2003/2004, the FD started a SF plantation in the cleared buffer zone surrounding the TWS. However, forest degradation has continued unabated, as has been reported by Belal (2013).

Social forestry has contributed greatly to conserving the forest area and people's livelihoods, particularly in the northern part of Bangladesh (Islam and Sato 2012). Similar approaches and SF systems were introduced in Teknaf to protect the forest area. This study investigates the status of SF programs and the level of forest destruction in 2013, 10 years after their establishment. Understanding the factors that caused differences in the effectiveness of these SF programs would be beneficial for successfully implementing appropriate future measurements for a regional conservation program.

10.2 Methodology

The state of the SF programs in 2013 was evaluated based on three methods of investigation: (1) interviews with forest officials, (2) field surveys, and (3) an analysis of satellite imagery.

10.2.1 Research Site and Timing of Field Studies

Surveys were conducted in the Teknaf and Shilkhali ranges after obtaining a research permit from the FD. Field surveys were conducted from September 9 to 17, 2013 and from March 4 to 9, 2014. The boundary of the visited ranges and zones (beats) of forest beat officers is given in Fig. 10.1.

10.2.2 Interviews

During the field surveys, interviews were conducted with officers and staff members at two ranges and seven offices of forest beat officers. A main questionnaire recorded the following items regarding the SF programs: year of establishment, area size, plant species, list of beneficiaries for each program, journal of each program, current conditions, and the program boundary. The exact location and boundary of each SF program were determined with the aid of a beat officer, forest guard, and headman. They also helped to trace the boundaries on satellite images (resolution: 1/10,000; GeoEye-1 2010). Subsequently, we visited each SF program in their company, and field measurements were performed.

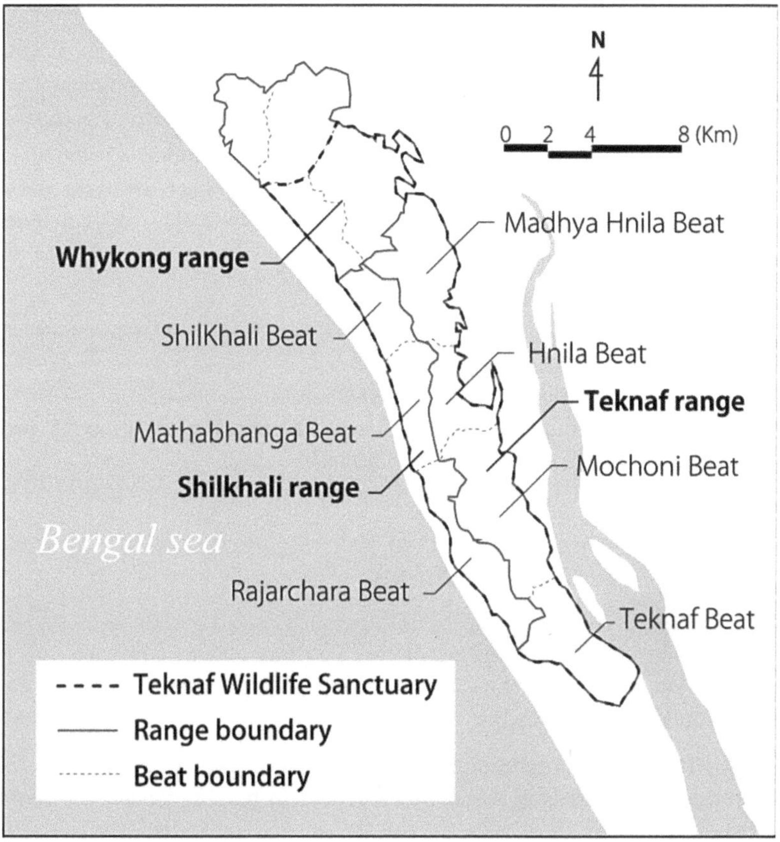

Fig. 10.1 Teknaf Wildlife Sanctuary showing different ranges and beats

10.2.3 Field Survey

A field survey was designed to study the current conditions of all of the SF programs. Because of the remoteness of the sites, it was not possible to measure several survey plots in each SF program; therefore, 1–3 representative survey plots were carefully selected. Each plot area had a 5-m radius. Tree height and diameter at breast height (DBH) were recorded for all trees with a DBH > 5 cm. Proportion of canopy cover, as assessed visually, and location (longitude and latitude measured using the Global Positioning System) were recorded for each plot. A photograph of the plot and a hemispherical photograph of the canopy from eye level were taken to calculate the canopy openness by software. Collected data were compared statistically.

10.2.4 Satellite Image Analysis

Polygonal data of each SF program area were created from satellite images using the geographic information system software TNTmips, version 2013 (MicroImages Inc., Lincoln, NE, USA) (GeoEye-1 2010). The vegetation index was calculated by masking agricultural fields and shading the image area for each set of the polygonal data. The normalized difference vegetation index (NDVI) was calculated as follows:

$$\text{NDVI} = \left(\text{near}-\text{infrared}\left(\text{NIR}\right)-\text{red spectral measurements}\right)/\left(\text{NIR}+\text{red spectral measurements}\right)$$

where NIR = 780–920 nm, and red = 655–690 nm.

From these images, the average NDVI was calculated across the SF program areas and the field survey plot areas (5-m radius).

10.2.5 Social Forestry Conditions

To determine the current conditions of the SF plots and to compare differences among the SF program areas, the following condition types were applied to each SF area according to its tree density:

Condition I: Tree density higher than the prescribed SF protocol
Condition $I_{\text{broad-leaf}}$: SF areas in which broad-leaved or bamboo species were planted and not cleared yet
Condition II: Tree density mostly the same as that designated by the SF protocol
Condition III_{cut}: SF area cleared before the harvesting year

10.2.6 Data Analysis

To validate the data obtained from the SF survey plots, the correlation between the survey data and the average NDVI values for the corresponding SF program areas was analyzed using a correlation matrix. Canopy openness was analyzed from the hemispherical photographs using CanopOn 2 v2.03c (CanopOn 2003; Takenaka). The conditions of the SF program areas were expressed as the increase of tree canopy cover and the increase in tree density since establishment. To evaluate the conditions of the SF program areas, the effects of tree density and thinning protocols were considered.

10.3 Results and Discussion

10.3.1 Number of SF Journals and Their Attribution

Seventy-one journals were collected from forest offices, 58 of which were from the Teknaf range and 13 were from the Shilkhali range. These journals included state forests and SF programs that have been devastated and thus had their contract with the beneficiaries terminated. However, the devastated forests are excluded from the list in Table 10.1. Thirty-seven active SF areas were identified in this study, 28 of which were located in the Teknaf range and 9 of which were in the Shilkhali range. The majority of the SF programs were established on the eastern side of the Teknaf peninsula. The implementation of the SF programs was started by the FD in 2003. Most of these programs were designed for short-rotation plantation forests (10-year harvest cycle) of *Acacia* spp. The proportions of planted species in the SF areas were calculated from the journals, and they are presented in Table 10.2. The FD not only planted fast-growing, non-native species (*Acacia* spp.) but also native species to increase the biodiversity in the forest. Several SF programs were found to be long-rotation forests (40-year harvest cycle) and were reforested for wild animal conservation. However, local people demanded short-rotation plants, such as *Acacia* spp., which can grow rapidly and be harvested early. As a result, SF programs with short-rotation species were more numerous than the others. Nevertheless, several other species were also found in SF program areas as well (Table 10.1). Hence, the social and biodiversity aspects of these SF programs are interesting future research areas.

10.3.2 Current Condition Types in 2013

The tree densities of the SF program areas are shown in Fig. 10.2. As per the SF regulations, there should be 2500 trees per ha on the plantations. Four years after the establishment of the SF programs, 50% of the trees should be thinned (leaving 1250 trees per ha after the first thinning). Then, 50% of the remaining trees should be thinned after 7 years (leaving 625 trees per ha after the second thinning). Finally, after completing the 10-year, short-rotation cycle, the remaining 625 trees per ha should be harvested.

Eight SF areas were classified as condition I, as they experienced delayed thinning operations (and were dominated by *Acacia* spp.). Fourteen SF areas were classified as condition II, as they were thinned by the beneficiaries as per the SF recommendations or for other unidentified reasons. Seven SF areas were classified as condition III_{cut}. Pictures of the different conditions are shown in Fig. 10.3. Other planted species include shegun (*Tectona grandis*), neem (*Azadirachta indica*), rain tree (*Samanea saman*), cane (bet) (*Calamus* sp.), chikrasi (*Chukrasia tabularis*), jarul (*Lagerstroemia speciosa*), sal (*Shorea robusta*), telsur (*Hopea odorata*), jamun (*Syzygium cumini*), bash (*Bambusa* sp.), boilum (*Anisoptera scaphula*),

Table 10.1 List of SF programs in the Teknaf and Shilkhali ranges and the corresponding plot field data, NDVI values, current conditions, and tree species

Attribution of social forest					Plot field data					NDVI				
Name of range					Proportion of canopy cover					Value of social forest area		Value of plot area (5 m radius)		
	The time in years since establishment					Openness of tree canopy								
	Name of beat	Establish year	Area (ha)	(y)	(%)	(%)	Density (the trees/ha)	Total of DBH area (cm2)	Species of the trees	Average	Standard deviation	Average	Standard deviation	Current condition type
Teknaf	Teknaf	2002–2003	15	10	60.0	32.7	635.7	1290.2	Akashmoni	38.2	12.5	40.0	6.7	II
	Teknaf	2002–2003	15	10	35.0	53.9			Cane, Chikrasi, Gorjon	41.7	10.7	39.0	5.1	Ib
	Teknaf	2003–2004	30	9	0.0				–	39.6	11.2	–	–	IIIc
	Teknaf	2004–2005	10	8	80.0	17.1	1271.5	3115.8	Akashmoni, Gamari	48.1	11.2	54.8	7.4	I
					75.0	19.7	1398.6	1974.3	Akashmoni, Gamari, Gorjon	–	–	50.4	6.7	
	Teknaf	2005–2006	20	7	0.0					37.2	9.3	–	–	IIIc
	Teknaf	2005–2006	20	6	5.0	38.9	1271.5	505.0	Agar, Akashmoni	38.2	10.6	34.5	5.5	II
					15.0	40.7	890.0	995.1	Mangium	–	–	40.0	6.2	
	Teknaf	2006–2007	20	6	70.0	16.4	1017.2	522.3	Teak	36.2	10.8	37.6	4.5	Ib
	Teknaf	2007–2008	4	5	95.0				Akashmoni	36.4	8.0	–	–	–
	Teknaf	2011–2012	40	1	0.0				Akashmoni	36.7	10.0	–	–	–
	Mochoni	2003–2004	30	9	25.0	28.6	635.7	2026.4	Akashmoni	41.9	10.6	23.8	5.0	II
	Mochoni	2003–2004	30	9	80.0	15.1	762.9	1707.0	Akashmoni	45.1	9.6	48.2	4.6	II
	Mochoni	2004–2005	20	8	75.0	19.5	1780.0	1473.8	Akashmoni	39.4	11.6	49.3	2.7	I
	Mochoni	2004–2005	20	8	10.0	51.8	635.7	1519.7	Akashmoni	41.9	10.6	46.3	7.6	II
	Mochoni	2005–2006	20	7	40.0	31.2	762.9	1359.9	Hybrid Mangium	42.5	10.6	48.4	9.9	II
	Middle Nhilla	2003–2004	20	8	35.0	28.4	1652.9	1699.5	Akashmoni	37.9	11.4	50.4	4.6	I
	Middle Nhilla	2003–2004	20	9	60.0	24.9	1017.2	1713.1	Hybrid	43.8	11.1	53.4	6.7	I
					80.0	29.9	890.0	2465.7	Sil koroi, Bahera	–	–	56.5	4.3	

	Middle Nhilla	2003–2004	5	9	5.0	71.3	254.3	2696.7	Gamari, Cane	34.4	11.0	32.7	5.8	Ib
	Middle Nhilla	2003–2004	5	9	30.0	30.0			Bamboo	40.1	9.9	38.1	3.8	Ib
	Middle Nhilla	2004–2005	20	8	40.0	32.2	1398.6	1377.4	Akashmoni	33.5	10.6	47.3	4.3	I
	Middle Nhilla	2005–2006	10	7	45.0	28.1	2034.3	1582.3	Akashmoni	42.8	10.8	52.8	3.9	I
					40.0	37.9	1398.6	1492.9	Akashmoni	–	–	–	–	
	Middle Nhilla	2007–2008	15	5	15.0	47.7	1271.5	628.9	Akashmoni	34.3	9.6	43.1	3.1	II
	Middle Nhilla	2007–2008	15	5	0.0					36.0	9.5	–	–	IIIc
	Middle Nhilla	2007–2008	25	5	40.0	36.7	1398.6	772.3	Akashmoni	34.7	9.4	34.1	4.2	II
	Nhilla	2002–2003	10	10	25.0	35.2	890.0	1096.9	Akashmoni	39.8	11.8	45.8	5.3	II
	Nhilla	2002–2003	10	10	35.0	24.9	635.7	1558.9	Akashmoni	44.6	11.1	46.0	4.3	II
	Nhilla	2004–2005	30	8	60.0	18.8	635.7	850.9	Akashmoni	40.9	11.9	42.6	5.3	II
	Nhilla	2005–2006	40	7	35.0	31.3	1017.2	844.1	Akashmoni	38.8	11.5	49.3	3.6	II
	Nhilla	2012–2013	30	0	0.0					–	–	–	–	–
Shilkhali	Rajar chora	2005–2006	20	7	35.0	31.7	890.0	1269.8	Akashmoni	41.1	13.2	52.8	7.6	II
	Rajar chora	2004–2005	10	8	85.0	21.6	890.0	2097.4	Sil koroi	42.0	13.3	50.0	4.6	Ib
					85.0	19.2	635.7	689.2	Gamari	–	–	43.3	7.9	Ib
					30.0	26.0	1144.3	1238.8	Akashmoni	–	–	49.9	7.0	II
	Mathabanga	2003–2004	20	9	85.0	15.2	1017.2	1671.1	Gamari	38.7	12.2	51.7	4.2	I
					85.0	16.2	1144.3	1471.7	Akashmoni	–	–	50.1	3.4	
	Mathabanga	2005–2006	20	7	75.0	14.2	2034.3	2084.2	Akashmoni	41.0	14.3	37.5	9.0	I
	Mathabanga	2011–2012	15	1	0.0					30.2	11.9	–	–	–
	Shilkhali	2003–2004	50	9	0.0					45.4	12.2	–	–	IIIc
	Shilkhali	2004–2005	20	8	0.0					39.7	12.6	–	–	IIIc
	Shilkhali	2005–2006	60	7	0.0				Akashmoni	32.1	11.1	–	–	IIIc
	Shilkhali	2007–2008	25	5	0.0					34.1	9.8	–	–	IIIc

Table 10.2 Proportion of planted tree species based on data extracted from the journals of the Teknaf and Shilkhali FD range offices

Bengali name	Scientific name	(%)
Gambar	*Gmelina* sp.	74
Akashmoni	*Acacia auriculiformis*	61
Hybrid akashmoni	*Acacia mangium*	58
Arjuna	*Terminalia arjuna*	53
Mahogany	*Swietenia mahogani*	51
Bahera	*Terminalia bellirica*	46
Amalaki	*Phyllanthus emblica*	45
Koroi	*Albizia* spp.	43
Horitaki	*Terminalia chebula*	26
Gorjon	*Dipterocarpus* sp.	25

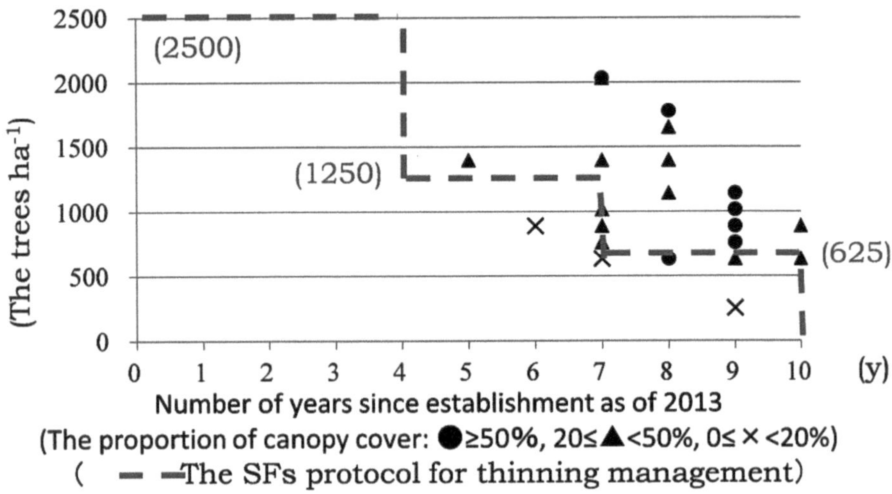

Fig. 10.2 Tree density and time (in years) since social establishment of *Acacia* spp. forests

kadam (*Neolamarckia cadamba*), dhakijam (*Syzygium firmum*), champak (*Michelia champaca*), sisu (*Dalbergia sissoo*), kantakari (*Solanum virginianum*), kaupha (*Garcinia cowa*), chulta (*Dillenia indica*), acacia (*Acacia decurrens*), jolpai (*Elaeocarpus robustus*), tetul (*Tamarindus indica*), dumur (*Ficus carica*), kathal (*Artocarpus heterophyllus*), ipil-ipil (*Leucaena leucocephala*), kola (*Musa* sp.), shimul (*Bombax ceiba*), chapalish (*Artocarpus chaplasha*), supari (*Areca catechu*), etc.

Condition I: Akashmoni (*Acacia auriculiformis*) SFs. Left; Middle Nhilla (2005-2006, 10ha) the year seventh and the tree density is 2,000 trees ha-1. Right; Mathabhanga (2003-2004, 20ha), the tree density is 1150 trees ha-1. These condition I SFs were with delayed thinning operation.

Condtion II: Akashmoni (*Acacia auriculiformis*) SFs. Left; Middle Nhilla (2007-2008, 15ha) the year fifth and the tree density is 1,270 trees ha-1. Right; Nhilla (2002-2003, 10ha), the year tenth and the tree density is 635 trees ha-1. The sunlight well through forest canopy and forest floor vegetation growth well too.

Condition $I_{broadleaf}$: Devastated SF. Left; Teknaf (2003-2004, 30ha) SF located close with town. It was happened when the year of national election in 2009. Right; Shilkhali (2005-2006, 60ha). FD officers claimed the inability to prevent thief activities who come from east side of peninsula through the mountain road.

Condition III_{cut}: Gamar (*Gmelina* sp.) is the most planted deciduous species. Left; Middle Nhilla (2003-2004, 5ha). Cane was planted onto the forest floor and former planted Gamar is remaining as standard. Right; Rajarchara (2004-2005, 10ha), the year eighth forest.

Fig. 10.3 Typical examples of stands for each condition of the SF areas

10.3.3 Correlation Analysis Between Variables

Correlations between different variables (the number of years since the establishment of the SF program, field survey data, and the NDVI) are shown in Table 10.3. The findings are described below.

Effectiveness of the Statistical Analysis Using Field Survey Data from 2013 and Satellite Imagery from 2010 There was a strong correlation between the proportion of canopy cover and the average NDVI of the survey plot (0.462, $P < 0.05$). Although there is a 3-year difference between the time of the field surveys and the time at which the satellite images were taken, the correlation matrix analysis provides a useful indication of the correlation between variables. The analysis showed some closed tree canopies in SF program areas where thinning operations were not conducted and some open canopies because of thinning operations.

Contribution of SF Programs to Reforestation There was a strong correlation between the number of years since the SF was established and the total DBH (0.476, $P < 0.05$) and between the number of years since establishment and the average NDVI of the SF areas (0.580, $P < 0.05$). Therefore, vegetation has increased within the SF programs. There was a significant correlation between the average NDVI of the SF areas and the proportion of canopy cover (0.363, $P < 0.01$), as well with tree canopy openness (−0.440, $P < 0.01$).

Effectiveness of the Value of the Proportion of Canopy Cover (Measured Visually) There was a strong, negative correlation between the proportion of canopy cover, which was measured visually, and the openness of the tree canopy (−0.795, $P < 0.05$), which was measured by hemispherical phótography. This result suggests that visual estimates of the proportion of canopy cover were accurate.

Table 10.3 Correlation matrix between variables: the time in years since SF area establishment

		I	II	III	IV	V	VI	VII
I	Years since establishment	–						
II	Proportion of canopy cover (%)	0.363*	–					
III	Canopy openness	−0.079	−0.795**	–				
IV	Tree density (trees ha^{-1})	−0.384*	0.153	−0.267	–			
V	Total DBH area (cm^2)	0.476**	0.267	−0.024	−0.002	–		
VI	Average NDVI of SF areas	0.580**	0.363*	−0.440*	0.031	0.421*	–	
VII	Average NDVI of survey plots	0.151	0.462**	−0.391*	0.260	0.241	0.410*	–

$^*p<0.05\%$, $^{**}p<0.01\%$

10.3.4 The Proportion of Canopy Cover in 2013

The proportion of canopy cover in 2013 and at the time of SF program establishment is shown in Fig. 10.4. The data were categorized into three groups based on the proportion of canopy cover. They include areas with less than 20, 20–50, and greater than 50% canopy cover. Figure 10.2 shows that 11 SF program areas had greater than 50% canopy cover, while the remaining 26 had less than 50%. This may be due to thinning activities that occurred in these 26 SF program areas. It was estimated that active tree felling seems to occur of those 26 SF program areas until 2013. As a result, differences were observed among the social forestry program areas regarding their current conditions.

Figure 10.3 indicates differences in the forest conditions of the different SF program areas. In the Teknaf range, the distribution of SF areas with conditions I and II was mixed. The density of the trees in condition II areas was mostly similar to the SF recommendation because of the thinning operations that were conducted in these areas. The BFD allowed some SF areas to perform thinning operations, while others were not because of poor growth and management. In such areas, destructive activities occurred. In the Teknaf beat, condition III_{cut} areas were located around the town of Teknaf. In the Shilkhali range, two condition III_{cut} areas were found in the Rajarchara beat. The area of the SF programs of this beat is rather small. In the Mathabhanga beat, three SF programs were found, all belonging to the condition I type. This suggests that the management of SF areas in this beat has been satisfactory. In contrast, all SF areas in the Shilkhali beat zone were condition III_{cut}, which is indicative of poor management practices. The mapping of these conditions may be helpful in identifying the risk level for deforestation and the need for local forest restoration.

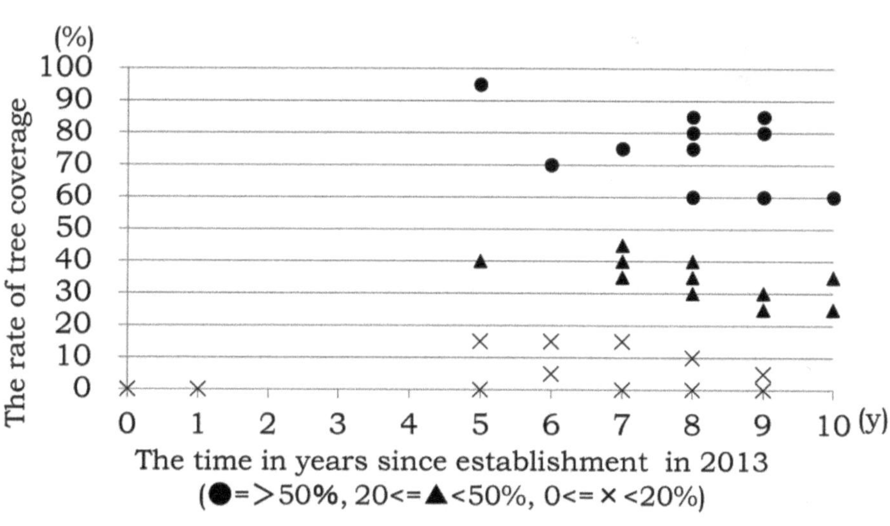

Fig. 10.4 Canopy cover and time (in years) since SF program establishment until 2013

10.3.5 Future Studies

Based on discussions with the officials of BFD and local stakeholders, some ideas were developed for a future study to conserve the forest in a sustainable manner.

Common Issue Many articles and studies have discussed the causes of deforestation and forest degradation in this area. Belal (2013) pointed out that the illegal collection of resources that occurred because of scarce patrolling may have resulted from a lack of logistical support and a shortage of staff. During the interviews conducted with the FD officers, some of them stated that beneficiaries and local people doubted that trees could be harvested successfully in SF areas, although the harvest was supposed to start in 2014. The beneficiaries and other people had already started unplanned thinning and tree removal operations before the designated harvest years. Another reason for forest deforestation and forest degradation was political clashes among the local people, which were numerous during 2009. Political unrest at that time created local conflicts that caused deforestation and forest degradation in many SFs. Some areas were burned, whereas others remained in good condition.

Local Issue There are several reasons for the poor management of the social forests on the western side of the Teknaf peninsula, including the occupations of the local people. Many local people in this area are engaged in betel leaf cultivation and fishing, which have higher income than other occupations. Therefore, they tend to be reluctant to engage in managing SF areas. A beat officer in the Rajarchara office claimed that a water shortage was a problem for nurturing and managing seedlings. Topographically, water catchments are small, and the water supply from streams and wells is not sufficient. The officers of the Shilkhali range and beat claimed that a lack of manpower prevented them from taking proper care of the SF areas and preventing burglaries in the SFs. Thieves come not only from the local area but also from the east side of the Teknaf peninsula. As a result, the conditions of the SF areas in the Shilkhali beat were worse than those of the others (Fig. 10.5).

The differences in the current conditions of different SF areas presented in this study can be used in future studies. These differences may result from social, political, topographical, and ecological factors. It is important to study forest conservation in cooperation with other researchers and to apply conservation measures locally. Currently, a comanagement approach, the Nishorgo Support Project, has been conducted to promote SF (Rahman 2011). It has also been stated that traditional, community-based forest management practices by indigenous people may be a useful guide for policymakers looking for ways to support sustainable forest management that involves local people. As mentioned, more traditional and local ways of conducting forestry strive to establish local forests for local people (Jashimuddin and Inoue 2012).

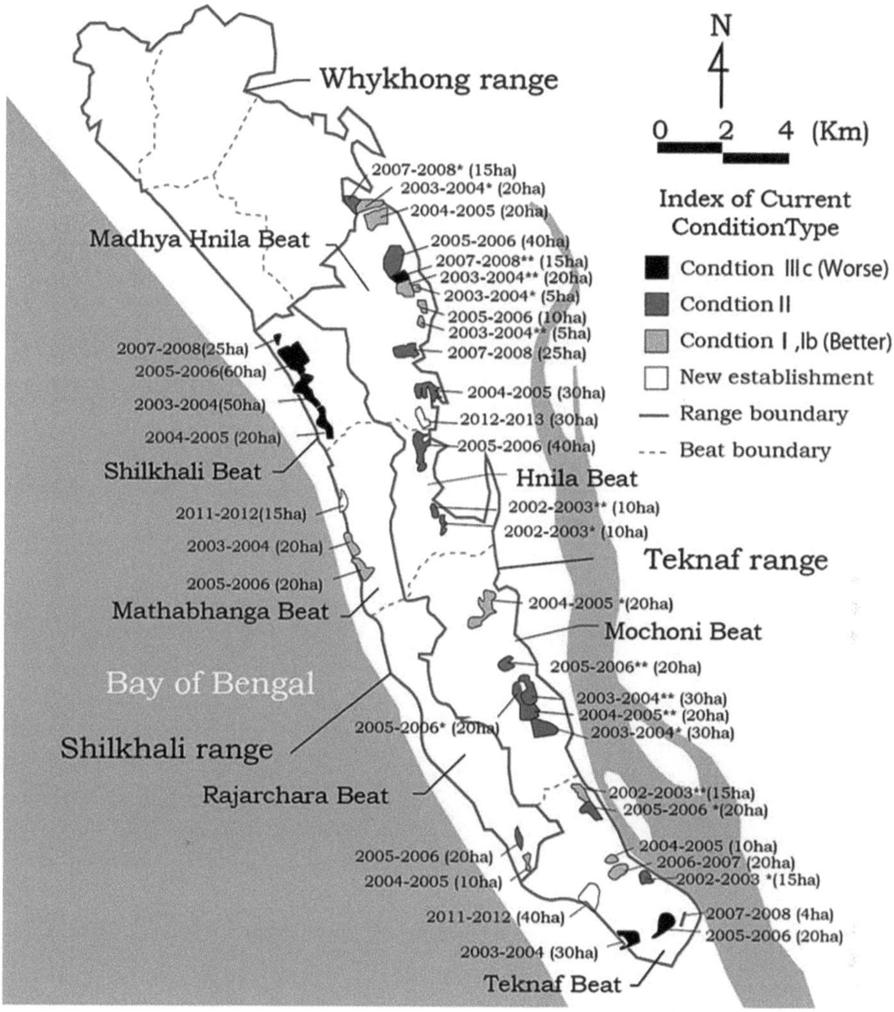

Fig. 10.5 Map of the of the SF areas of the Teknaf and Shilkhali ranges in 2013

10.4 Conclusions

In the Teknaf and Shilkhali ranges, 37 active social forests were identified having proper management practices in 2013, 10 years after SF started in this area. However, current conditions vary among these SF areas. *Gmelina arborea* and *Acacia* spp. were the most dominant tree species in the studied social forests. Five broad-leaved social forests are also identified, while seven social forests are found to be cleared. It is found that tree felling occurred in approximately 70% of the social forests. Thinning is delayed in eight social forests, whereas 11 are thinned properly.

References

Belal MAH (2013) Effectiveness of co-management committees in Teknaf Wildlife Sanctuary. In: Fox J, Mustafa MG, Bushley Brennan SM, Durand L (eds) connecting communities and conservation: co-management initiatives implemented by IPAC in wetland and forests of Bangladesh. USAID-Nishorgo network, Dhaka

Islam KK, Sato N (2012) Participatory forestry in Bangladesh: has it helped to increase the livelihoods of Sal forests dependent people. South ForJ For Sci 74(2):89–101

Jashimuddin M, Inoue M (2012) Community forestry for sustainable forest management: experiences from Bangladesh and policy recommendations. FORMATH 11(0):133–166

Rahman MA (2011) Deforestation and forest conservation in a Tanchangya Community. In: Fox J et al (eds) Rural livelihoods and protected landscapes: co-management in the wetlands and forests of Bangladesh. East-West Center, Honolulu, pp 118–132

Salam MA, Noguchi T (2005) On sustainable development of social forestry in Bangladesh: experiences from Sal (*Shorea robusta*) forests. Environ Dev Sustain 7:209–227

Chapter 11
Actors and Their Power in Social Forest Management

K.K. Islam and N. Sato

Abstract Social forestry (SF) has the potential to improve rural livelihoods and alleviate poverty among forest-dependent people. It can also help to protect forests against encroachment and illegal felling. Many actors are involved in the implementation and execution of SF, which is inherently political because of competing access to and control over forests embedded within social and power relations. Consequently, SF entails an emblematic struggle for dominance and power between diverse actors. A study was undertaken in Teknaf peninsula, which contains highly degraded forests, to examine the extent of power and livelihood assets of actors engaged in social forest management. Seventeen actors were identified in relation to SFs in Teknaf, which contributed to the program's complex and imbalanced power dynamics. The forest administration retained the most power at each level and played a dominant role in decision-making and other management activities. The analysis of livelihoods revealed that the SF program has had positive impacts on the livelihood capital of beneficiaries, indicating that it is an appropriate managerial approach for improving livelihoods within local communities while simultaneously protecting forest cover. Based on the study's findings, decentralization of power and a reduction of the actors involved in SF are highly recommended. Further, there is a need for more intensive training and the development of appropriate technologies for tree-crop cultivation in Teknaf SF.

Keywords Actors • Social forestry • Forest management • Beneficiaries • Livelihood

K.K. Islam (✉)
Department of Agroforestry, Faculty of Agriculture, Bangladesh Agricultural University, Mymensingh, Bangladesh
e-mail: kamrulbau@gmail.com

N. Sato
Forest Policy Laboratory, Faculty of Agriculture, Kyushu University, Fukuoka, Japan
e-mail: sato.noriko.842@m.kyushu-u.ac.jp

M. Tani, M.A. Rahman (eds.), *Deforestation in the Teknaf Peninsula of Bangladesh*, DOI 10.1007/978-981-10-5475-4_11

11.1 Introduction

In recent decades, there has been growing international recognition of the shift in forest resource management away from top-down, traditional management toward a more people-oriented approach that incorporates local communities and all stakeholders in decision-making (Brown 2009; Ribot 2004). In addition, decentralization in forest management policies started as a response to institutional malfunction, and during the last decade, institutional changes such as social forest management paradigms have become major policy trends in many of the world's developing countries (Prasad and Kant 2003; Ribot 2004; Rosyadi et al. 2005; Shahbaz 2009; Siry et al. 2005). Many countries have already developed, or are developing, changes to state laws and policies that institutionalize social forest management approaches. The main initiative of social forestry (SF) is to develop joint management between local communities and forest departments (representing the state) to manage and conserve forests on the basis of a friendly relationship and trust. Thus, the new management system is a way to improve cooperation and to reduce insecurity in the relationship between the two entities and other actors (Gardener et al. 2001; Shahbaz 2009). A theory of the involvement of local people in the management of common natural resources supports social forest management's potential for attaining ecological sustainability, economic efficiency, and social equity (Agrawal and Ostrom 2008; Bowler et al. 2012; Ostrom 1990). However, a literature review also showed that joint management between forest departments and other actors can initiate new conflicts or cause old ones to worsen (Castro and Nielsen 2001). Therefore, the transformation of new forest management schemes into a people-oriented approach is not easy; in fact, it is a major challenge in many countries (Fisher 2007; Islam and Sato 2012a, b; Nygren 2005).

Social forest management is characterized by many stakeholders because of the economic, ecological, and social functions and values that forests deliver. In addition to local communities, other groups at the regional, national, and international levels also have impacts on local communities' access to forests (Peluso et al. 1994). Theoretically, four broad categories of actors are involved in people-oriented forestry: the state, the local community, the private sector, and the donors (Hobley 2004; Sharma and Acharya 2004). All four actors are important in forest management, and their cooperation is needed for sustainable forest management. Often, the state has the predominant role (Barrow et al. 2002; Devkota 2010; Krott 2005).

In Bangladesh, social forest management started in 1981/1982 with the funding of donor agencies (Islam and Sato 2012b). Until then, SF was regarded as a government-controlled and donor-funded enterprise in Bangladesh (Islam and Sato 2012a; Islam et al. 2012). It has also been regarded as political in nature because of its contested access to, and control over, forests in social and power relations. Bangladesh forest ecosystems are composed of the hill, sal (*Shorea robusta*), and mangrove forests, and these forests display great richness and diversity of cultures and people, geographical features, flora, and fauna. SF covers both the hill and sal forests, while mangrove forests are restricted to a United Nations Educational,

Scientific, and Cultural Organization World Heritage Site. Evergreen and semi-evergreen hill forests (0.67 million ha) of Bangladesh are situated at the eastern boarder of the country, and they are considered to be ecologically and economically important (Islam and Sato 2012a, 2013; Muhammed et al. 2008). Forests on the Teknaf peninsula are mainly composed of evergreen hilly species, and they are rich in biodiversity. Forest management approaches have created great diversity in the social relationships and interests among actors. Therefore, SF initiated an emblematic struggle between diverse kinds of actors in terms of their dominance and power relations. Nevertheless, power has played a progressively important role in forest policy analysis since the implementation of SF on the Teknaf peninsula. SF views local communities as the main actors who use forests in different ways for a wide variety of reasons (Devkota 2010; Islam and Sato 2012a; Muhammad et al. 2008). Thus, there is an immediate need to pinpoint key issues relating to different actors, their power and interests in social SF activities, and their influence on the policy cycle in Teknaf and Bangladesh. Therefore, this chapter tries to identify the powerful actors and the extent to which they affect the decision-making and empowering process of social forest management on the Teknaf peninsula.

11.2 Theory and Concept of Actor Power

The notion of power has become progressively less important in forest policy analysis over the last few decades. Power generally refers to the ability to impose one's will or advance one's own interest (West 2004). It is also anticipated that participation can overcome power imbalances by involving all actors in a process that meets their interests, and this study addresses the hypothesis that power may alter the outcome of participatory behavior. Maryudi (2011) defines actor-centered power as "a social relationship in which actor 'A' alters the behavior of actor 'B' without recognizing B's will." Actor-centered power influences participatory forest management via the most powerful actors and their diversified interests. Therefore, the research for this study relies on Webber's (1964) theory of power against resistance (coercion and incentives) with a new dimension of power without resistance (trust) (Maryudi 2011); thus, power is clearly composed of coercion, incentives, and trust elements. According to Webber (1964), acts of resistance could bring limited power to social relations, and the resistance could be broken forcefully through coercion or softly by incentives. In addition to Webber's thinking, there is the possibility that power relations can be present without resistance, i.e., trust. Thus, the elements of actor-centered power consist of coercion, incentives, and trust (Maryudi 2011; Webber 1964) (Fig. 11.1).

Simply stated, trust is a power element through which the subordinate changes his/her behavior by accepting the potentate's information (Devkota 2010; Itubo 2011). Power is exercised by the use of information. Coercion is the practice whereby an individual or a group of people is forced by another party to behave involuntarily. This is made possible by either action or inaction (Itubo 2011; Krott 2005).

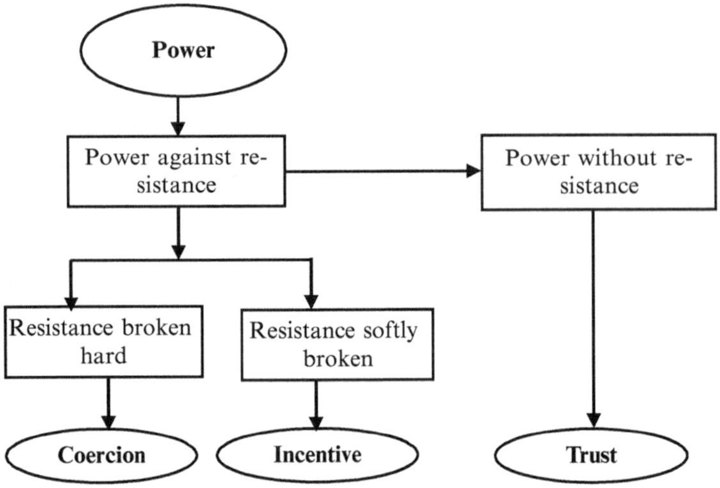

Fig. 11.1 Elements of actor-centered power

However, incentives are financial or nonfinancial factors that change a subordinate's behavior via motivation. Here, motivation is the initiation of a goal-oriented attitude and the expectation of benefits that encourage people to change their behavior. Thus, the actor-centered power conception is regarded as more useful for the analysis of power in the case of forest management and policy issues.

11.3 Research Approach

In Bangladesh, evergreen and semievergreen forests cover 670,000 ha, and these forests are owned by the Bangladesh Forest Department (FD) (FD 2014; Islam and Sato 2012a). The tropical evergreen and semievergreen forests extend over the southern part of the country, consisting of Chittagong, Cox's Bazar, Chittagong Hill Tracts, and Sylhet districts. The study was conducted on the Teknaf peninsula under tropical evergreen and semievergreen forests in the Cox's Bazar area of Bangladesh. This study collected quantitative and qualitative data, and quantitative data were collected through a semi-structured questionnaire survey. To obtain qualitative data, this study used discussions with FD staff and local people, a focus group discussion, personal observations, and a literature review. SF programs were mainly implemented with the active guidance of forest beat offices (the lowest FD administrative office). Teknaf forests consist of 11 beat offices, and the study covered seven beats. This study randomly selected 40 members from the Teknaf SF area. During field visits, actors were asked about their views of other actors, and this study tried to cover all of the SF actors listed in the Results section. The interview questionnaire was pretested and improved before conducting the final interview, and a research team, which consisted of seven members, was involved in data collection in the Teknaf area from March to September 2014.

11.3.1 Description of the SF Programs

In the SF programs, each member is allocated 1 ha of degraded forestland for a SF plantation with a 10-year rotation cycle. Each farmer can continue for up to three rotation cycles (30 years) if he/she meets the SF criteria. Fast-growing firewood tree species (e.g., *Acacia auriculiformis*) are selected for plantation, with a spacing of 2 m × 2 m (a total of 2500 tree per ha). After 4 years, 50% of the standing trees are thinned (first thinning), and this technique is repeated after 7 years (second thinning). The remaining 625 (approximately) trees are finally harvested at the end of the 10-year cycle. The FD and members share the benefit of the second thinning, and final tree harvest outputs at a ratio of 45:45, while the remaining 10% benefit, which is called the tree-farming fund, is saved for the next rotation of the tree plantations. The farmer could grow annual crops in association with the trees at any time during the 10-year rotation cycle, and the crops, together with the first thinning benefits, are granted solely to the farmer. These types of people-oriented programs have gained popularity throughout Bangladesh.

11.3.2 Analytical Frameworks

The study's questionnaires collected information that identifies power status, based on the power elements of trust, coercion, and incentives, of actors within the SF networks. The study covered every actor and asked each actor about their judgment regarding the power elements coercion, incentives, and trust of the other actors. The findings of the three power elements were categorized using a scale from 0 to 3 (0 = no power and 3 = complete/highest power), and each actor gave a specific power dimension of the other actors in the networks. Finally, the average value of each actor was coded to determine the most powerful and least powerful actors; we applied a dominance degree analysis model (Schmidt 2000) to categorize the most powerful actors. The analytical techniques used to differentiate the most powerful actors from the least powerful ones were developed by Schusser (2013), in which an individual's relative power (X_i) and dominance degree (D_i) were used to identify the powerful groups in the SF networks using the following formulas (Jonas and Pfiterer 2010; Maryudi 2011):

$$h_i = \frac{Xi}{\sum_{i=1}^{n} Xi} \quad CR_r = \sum_{r}^{i=0} hi \quad D_i = \frac{\frac{CRi}{i}}{\frac{1-CRi}{n-i}}$$

and

$$r \leq n$$

Here, n is the total number of actors identified; X_i is the sum of the answers per actor; h_i is the ratio of the power per actor and the power element (i), with $0 > h_i \geq 1$; and r is the number of powerful actors considered.

The above formulas were used to analyze the dominance degree (D_i), which is calculated by first sorting the data from the highest value to the lowest. Subsequently, all values (X_i) are related by calculating the relationship (h_i) between the power element per actor (i) and the sum of the power elements for all actors. After calculating the h_i of each actor under each power element, the cumulative accumulated value (CR_i) of each actor in the network can be calculated. Finally, the dominance degree (D_i) can be calculated using the above formulas.

11.4 Results

The study first identified the actors involved in the SF networks, and it eventually determined those who were deemed to be the most powerful using the simple 0–3 scaling system. The study identified 20 actors, and in each SF network, the individual relative power (X_i) and dominance degree (D_i) calculations were used to identify and score the powerful actors. Teknaf SF networks showed that the forest administrations, particularly the beat officer, were the most powerful actors (i.e., the Ministry of Forests and Environment and the Park Management Authority), including forest administrations, had the highest power in the SF networks, and thus controlled the SF programs in all three domains (trust, incentives, and coercion) of power dynamics (Table 11.1). Table 11.1 also shows all of the government actors.

The dominance degree (D_i) calculation gave a sequence (powerful to less powerful) of the dominant actors, and it graphically showed the most dominant actors in the trust power element according to their D_i value in the SF networks (Fig. 11.2). In Teknaf SF, the most dominant actor was the beat officer, followed by the divisional forest officer (Fig. 11.2) (In Fig. 11.2, the powerful actors are presented in chronological order, from left to right).

Concerning the incentives element of the power analysis, the results showed that many actors provided incentives, but the dominant one was the forest administration, which included different projects funded by the donor agencies. In contrast, the coercion element of the power analysis clearly showed that only the forest administration had the most power over the other actors (Table 11.1).

11.5 Discussion

The power analysis provides the basis for the quantitative analysis that focuses on how the identified, powerful actors shape and accrue their power (Maryudi 2011). In every instance of the Teknaf SF program, the forest administration showed the highest scores in the quantitative analysis of the actors' networks (Table 11.1).

Table 11.1 Summary of the power analysis of Teknaf SF

Actor category	Power dimension (X_i)			Actors in the networks
	Trust	Incentive	Coercion	
Forest Department	28	8	4	Divisional forest officer
	20	9	3	Range officer
	35	12	5	Beat officer
Donor	20	6	2	Upazila Afforestation and Nursery Development Project
	19	12	1	Costal Greenbelt Project
	27	12	1	Forestry Sector Project
	26	11	1	Nishorgo Support Project
State Ministry (central level)	24	12	1	Ministry of Forests and Environment
Forest Department (central level)	25	11	0	Wildlife Management and Nature Conservation Division
Private sector	8	4	0	Sawmillers
	7	4	0	Brickfields
Leader	11	9	0	Local political leaders
	17	15	0	Union Parishad leaders
Social Forest Association	16	10	4	Local social forest committee
Development organizations	18	11	0	BRAC, CREL (NGOs)
Individual	13	4	3	Headman
	4	6	0	Encroachers

BRAC Bangladesh Rural Advancement Committee, *CREL* Climate-Resilient Ecosystems and Livelihoods

The forest administration, especially the beat officer, remains the most powerful actor in the trust (35 points), incentives (12 points), and coercion (5 points) elements of the power analysis. This is because the beat officer is officially responsible for selecting the SF members and evaluating the SF programs and the benefit-sharing process (Islam and Sato 2012a). However, the donor-funded projects, such as the Forestry Sector Project and the Coastal Green Belt Project, provided the maximum incentives, as well as being trusted in the network. The local SF committee was also shown to have a certain degree of power, although the results suggest that their power appears to be limited by the amount of trust placed in them by the other actors in the networks. In an analysis of actor power in a community forestry study of Nepal, Devkota (2011) found that forest administrations had the highest level of power in the trust, incentives, and coercion elements. Similarly, forest administration remains one of the most powerful actors not only in coercion strategies but also in providing incentives and being trusted, as determined by an analysis of actors' network dynamics in community forestry programs in Indonesia, Namibia, Albania, China, the Philippines, and Cameroon (Devkota 2011; Itubo 2011; Krott et al. 2014; Maryudi 2011).

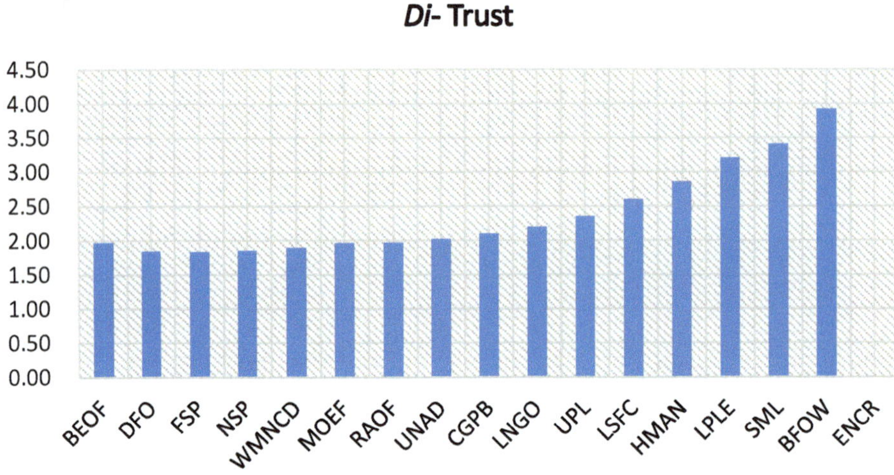

Fig. 11.2 Assessment of the degree of dominance factor of trust for Teknaf SF. (*BEOF* beat officer; *DFO* divisional forest officer; *FSP* forestry sector project; *NSP* Nishorgo Support Project; *WMNCD* Wildlife Management and Nature Conservation Division; *MOEF* Ministry of Forests and Environment; *RAOF* range officer; *UAND* Upazila Afforestation and Nursery Development Project; *CGPB* Coastal Greenbelt Project; *LNGO* local nongovernment organizations; *UPL* union Parishad leader; *LSFC* local social forestry committee; *HMAN* headman; *LPLE* local political leader; *SML* sawmillers; *BFOW* brickfield owner; *ENCR* encroachers)

The actor network survey results of the study sites were obtained using the rule that states that each individual actor who is a part of a powerful actor, with regard to at least one power element, is considered to be a part of the most powerful actor (Schusser 2013). The dominance degree (D_i) values clearly show the point of separation between the group of powerful actors, and less powerful actors were found to have the highest dominance values. Based on the dominance degree value, the most powerful actor was identified (Schusser 2013), and Fig. 11.2 clearly shows the dominant actors in their networks, considering the trust element. Schusser (2013) used dominance degree calculation systems to determine the power dynamics in community forestry programs in Cameroon. However, a ranking system to determine the effect of community forestry on the forest using quantitative data was also applied by Coleman and Fleischman (2012). In SF research, a number of scientists stated that power decentralization was rarely followed by genuine power devolution to the local members (Islam and Sato 2012a; Larson and Ribot 2007; Ribot 2004). The more powerful actors have a tendency to manipulate such outcomes to maintain their position in the forest management regime (Shackleton et al. 2002). Thus, local forest users were the least powerful actors in the decision-making process of any people-oriented forest management approaches (Islam and Sato 2012a, 2013), although the SF policy clearly outlined the decentralization and devolution of power to local level.

11.6 Conclusions and Policy Implications

SF programs in Bangladesh have, without doubt, introduced a new interpretation of forest management by employing an approach that includes local people together with rural development and resource conservation. This approach also includes many actors because of the economic, ecological, and social functions and values that forests deliver. In this study site, the results showed that the FD was the most powerful and influential actor in SF. The FD is the most powerful actor in all three power elements of the actors' power analysis, and it gained numerous power features through the bureaucratic forest management policies of Bangladesh. In this study, the actors' power analysis demonstrated that there is an immediate need to empower the local people and their committees, and together with this empowerment issue, the study also argued for some other important steps that can improve the power imbalance among the actors, as well as the programs as a whole. These steps are:

- Institution building and empowering local-level organizations (e.g., forest user groups in India and Nepal), in which the local people/community play the central role in governing all of their development initiatives. This means that the Bangladesh Government needs to provide appropriate facilities to the local communities to strengthen the local-level committee/institutions and that the FD should be the facilitator.
- It is necessary to develop clear decision-making power and rights regarding the use of forest resources. Forest use rights of the local people who have been using the forests for a long time should also be recognized.
- Central governments also need to establish strict monitoring and controlling systems over all brickfields, sawmills, and timber merchants situated in the Teknaf forest area so that they cannot use illegal timber to control the timber market.

References

Agrawal A, Ostrom E (2008) Decentralization and community-based forestry: learning from experience. In: Web EL, Shivakoti GP (eds) Decentralization, forests and rural communities: policy outcomes in South and Southeast Asia. Sage Publication, New Delhi, p 44

Barrow E, Clarke J, Grundy I, Jones KR, Tessema Y (2002) Analysis of stakeholder based natural resource management: creating space for local people to participate and benefit? Natural resource perspectives. University of California Press, Berkeley, p 320

Bowler DE, Buyung-Ali LM, Healey JR, Jones JPG, Knight TM, Pullin AS (2012) Does community forest management provide global environmental benefits and improve local welfare? Front Ecol Environ 10:29–36

Brown FP (2009) Participatory forest management (PFM) discourse in South Africa: ecological moderation in the developing world [dissertation]. University of KwaZulu-Natal, Durban

Castro AP, Nielsen E (2001) Indigenous people and co-management: implication for conflict management. Environ Sci Policy 4:229–239

Coleman EA, Fleischman FD (2012) Comparing forest decentralization and local institutional change in Bolivia, Kenya, Mexico, and Uganda. World Dev 40(4):836–849
Devkota M (2010) Interests and power as drivers of community forestry: a case study of Nepal [dissertation]. The University of Gottingen, Gottingen
Devkota M (2011) Dimension of community-based forest management in the Philippines. Int For Rev 8(4):377–394
FD (Forest Department) (2014) Protected areas of Bangladesh. Ministry of Environment and Forest, Government of Bangladesh [Internet]. [last cited 13 March 2015]. Available from: http://www.bforest.gov.bd/
Fisher RJ (2007) Participatory forest management, sustainable livelihoods and poverty reduction: experiences from community forestry in Asia. Paper presented at: international conference on participatory forest management, biodiversity and livelihood in Africa, 17–19 March. Addis Abba
Gardener A, De-Marco AJ, Asanga AC (2001) A conservation partnership: community forestry at Kilum- ljim, Cameroon. Rural development forestry network. Network Paper 25h. Overseas Development Institute
Hobley M (2004) Players in the sector-civil society, private sector and donor agencies [Internet]. [cited 12 June 2014]. Available from: http://www.cambodia-forestsector.net/docs-part2.htm
Islam KK, Sato N (2012a) Participatory forestry in Bangladesh: has it helped to increase the livelihoods of Sal forests dependent people. South For J For Sci 74(2):89–101
Islam KK, Sato N (2012b) Deforestation, land conversion and illegal logging in Bangladesh: the case of the Sal forests. iForest 5:171–178
Islam KK, Sato N (2013) Protected Sal forests and livelihoods of ethnic minority: experience from Bangladesh. Sustain For 32(4):412–436
Islam KK, Ullah MO, Hoogstra M, Sato N (2012) Economic contribution of participatory agroforestry program to poverty alleviation: a case from Sal forests, Bangladesh. J For Res 23(2):323–332
Itubo AF (2011) Power dynamics within community forestry: examples from Kenya. MSc, dissertation, Gottingen University
Jonas N, Pfisterer L. (2010) Measurement of concentration and market power. In: seminar report, the University of Kaiserslautern, Germany [Internet]. [cited 12 June 2014]. Available from: http://www.wiwi.unikl.de/dekanat/blank/Segelseminar2010/11/Marktmacht.pdf
Krott M (2005) Forest policy analysis. Springer, Dordrecht
Krott M, Bader A, Schusser C, Devkota R, Maryudi A (2014) Actor-centered power: the driving force in decentralized community based forest governance. Forest Policy and Econ 49:34–42
Larson AM, Ribot JC (2007) The poverty of forest policy: double standards on an uneven playing field. Sustain Sci 2(2):189–204
Maryudi A (2011) The contesting aspiration in the forests: actors, interests and power in community forestry in Java, Indonesia [dissertation]. The University of Gottingen, Gottingen
Muhammed N, Koike M, Haque F, Miah MD (2008) Quantitative assessment of people oriented forestry in Bangladesh: a case study from Tangail Forest Division. J Environ Manag 88(1):83–92
Nygren A (2005) Community-based forest management within the context of institutional decentralization in Honduras. World Dev 33(4):639–655
Ostrom E (1990) Governing the commons: the evolution of institutions for collective action. Cambridge University Press, New York
Parasad R, Kant S (2003) Institution, forest management, and sustainable human development-experiences from India. Environ Dev Sustain 5:353–367
Peluso NL, Tumer M, Fortman L (1994) Introducing community forestry: annotated listing of tropics and readings. FAO, UN, Rome
Ribot JC (2004) Waiting for democracy: the politics of choice in natural resource decentralization. World Resource Institute, Washington, DC, p 140

Rosyadi S, Birner R, Zeller M (2005) Creating political capital to promote devolution in the forestry sector – a case study of the forest communities in Banyumas district, Central Java, Indonesia. Forest Policy Econ 7:213–226
Schmidt VA (2000) Democracy and discourse in an integrating Europe and a globalizing world. Security 24(4):5–38
Schusser C (2013) Comparative analysis of community forestry: theoretical and methodological requirements [dissertation]. the University of Gottingen, Gottingen
Shackleton S, Campbell B, Wollenberg E, Edmunds D (2002) Devolution and community based natural resource management: creating space for local people to participate and benefit? Natural resource. Perspective 76:1–6
Shahbaz B (2009) Dilemmas in participatory forest management in northwest Pakistan: a livelihoods perspective. Hum Geograph Ser 25:15–16
Sharma NN, Acharya B (2004) Good governance in Nepal's community forestry: translating concepts into action. In: Kanel KR, Mathema P, Kandel BR, Niraula DR, Sharma AR, Gautam M, (eds) Proceeding of the 4th national workshop on community forestry, 2004, August 4–6, Kathmandu
Siry JP, Frederick WC, Ahmed MR (2005) Sustainable forest management: global trends and opportunities. Forest Policy Econ 7:551–561
Webber M (1964) Basic concept of sociology. Free Press of Glencoe, New York
West WF (2004) Formal procedures, informal processes, accountability, and responsiveness in bureaucratic policy making: an institutional policy analysis. Public Adm rev 64(1):66–80

Chapter 12
The Homestead as a Production System

Md. Abiar Rahman

Abstract From ancient times, family-managed homesteads have supplied various products. In Bangladesh, there are diverse homestead production systems, which vary according to factors such as location, farm categories, local preferences, and ecological conditions. The Teknaf peninsula has a particular ecology that influences homestead production. We grouped homesteads based on farm-holding categories and evaluated the socioeconomic conditions, components, tree diversity and arrangements, and contributions of homesteads to households' annual incomes. Households within a total of 180 homesteads, grouped into five categories (landless, marginal, small, medium, and large), were surveyed using a structured interview schedule. The homesteads were also examined and monitored to assess their tree diversity and performance. Homesteads belonging to households in the large farm category were wider and were established earlier than those within the other categories. Among homestead components, fruit trees were predominant across all of the categories, followed by livestock and poultry and vegetables. Very few homesteads had medicinal plants. Interestingly, households were not inclined to plant timber species. This may be because they have access to easily available and cheap timber from the forests. Betel nut, mango, jackfruit, and coconut are some most dominant tree species. Larger homesteads had a greater number of trees and higher tree diversity. Accordingly, incomes derived from homesteads were significantly higher for those households in the large farm category, with incomes showing a gradual decrease corresponding to a reduction in farm size. However, while the percentage contribution of homestead incomes to annual incomes was the highest for households in the large farms category, the second highest contribution of homestead incomes was found for households in the marginal farm category, indicating that this group tended and efficiently managed more plant species than those belonging to other categories. We concluded that the homestead production system could be improved through a design accommodating many components for maximizing outputs and income generation.

M.A. Rahman (✉)
Department of Agroforestry and Environment, Faculty of Agriculture, Bangabandhu Sheikh Mujibur Rahman Agricultural University, Gazipur, Bangladesh
e-mail: abiar@bsmrau.edu.bd

M. Tani, M.A. Rahman (eds.), *Deforestation in the Teknaf Peninsula of Bangladesh*, DOI 10.1007/978-981-10-5475-4_12

Keywords Farm holdings • Socioeconomic conditions • Production components • Tree diversity • Income

12.1 Introduction

A homestead is both a family dwelling and a production unit where plants, animals, and fish are found with continuous interactions occurring among the components (Foysal et al. 2013). It is a traditional system, which is mostly practiced in tropical and subtropical regions as a center of livelihood and subsistence. According to Fernandes and Nair (1990), a homestead deals with the deliberate management of multipurpose trees and shrubs in association with various components such as agricultural crops, livestock, poultry, and fisheries within the compound of an individual house. In general, the homestead production system is managed primarily by family members and is complex and intensive. Thus, it is an economically efficient, ecologically sound, and biologically sustainable production system.

As a low-cost production system, it supplies food and nutrition, increases household income, and provides various services and functions. It helps in protecting the ecological habitat (Landauer and Brazil 1990; Marsh 1996). Generally, household members consume most of the products they produce, and the rests are sold in the market. Halladay and Gilmour (1995) noted that multiple functions related to daily human life can be obtained from a homestead and maintain a good ecosystem through all living organisms and their physical environment. A well-managed homestead provides food, nutrients, domestic energy, fodder, medicinal products, timber, and a pleasant environment for the families that dwell in it. Socioeconomic benefits at individual as well as community levels in the rural area can be ensured as a wide range of plants are grown in a homestead by providing shade, shelter, recreation, and agroecological balance (Roy et al. 1996). The ecological benefits from a homestead are regulated by soil, water, nutrients, and biodiversity conservation (Kabir and Webb 2008).

Homestead is an ancient practice, which is found throughout Bangladesh. Based on current conditions in the country, homesteads are an appropriate option for resource-poor farmers, who benefit from crop production and earn immediate cash incomes while receiving long-run benefits from the plants. The general pattern of homesteads in Bangladesh entails a multistoried structure with various plant species. The biodiversity in a tropical home garden is more rich and efficient than tropical rain forests in terms of their production, protection, and value. A typical homestead in Bangladesh provides a valuable opportunity for a household to engage in a number of economic activities undertaken in and around it (Rahman et al. 2016; Chowdhury and Satter 1992), which contribute approximately 70% of fruit, 40% of vegetables, 70% of timber, and 90% of firewood and bamboo produced (Abedin and Quddus 1988; Millat-E-Mustafa and Teklehaimanot 1996). Some other products include brooms, handicrafts, and shade-providing items, as well as products with ornamental, ceremonial, environmental, and aesthetic uses. Generally, the homestead production is managed by the household family members particularly women that empowers them by decision-making and new income-generating opportunities (Iannotti et al. 2009). The on-farm research division of the Bangladesh Agricultural Research Institute developed few homestead vegetable production models and reported a significant increase in the productivity and nutrition-supplying capacity of the householders (Ali et al. 2006). Due to increasing population and fragmentation of families, agricultural land is decreasing, but homestead number is increasing. Even a landless farmer has a homestead which offers an effective survival strategy for farmers, as it enables them to attain food and income security (Miah and Hussain 2001). However, the distribution and composition of components and benefits vary between ecosystems found in Bangladesh.

The Teknaf peninsula has diversified ecosystems where forest, marine, and agriculture are found in a narrow space. However, the resources are degraded due to overexploitation especially of forest resources. Most of the people depend on the forests for livelihood, income generation, and collection of forest and non-timber forest products. A well-managed homestead has the potentiality to supply various products in meeting the daily needs of rural people and can conserve biodiversity, which has not yet been studied in the Teknaf peninsula. This chapter assesses the homestead production system in the Teknaf peninsula in terms of its contribution to species composition, diversity, and income generation.

12.2 Methodology

Baharchara union of Teknaf Upazila in Cox's Bazaar district was selected for the study (Fig. 12.1). A total of 100 sample homesteads were selected for data collection. The homesteads were selected from the following farm categories: landless (<0.2 ha), marginal (0.21–0.5 ha), small (0.51–1.0 ha), medium (1.1–2.0 ha), and large (> 2.0 ha) (Abedin and Quddus 1988; Zaman et al. 2010). It was assumed in categorizing homestead that the resources and wealth of a household are influenced by farm size that plays a vital role in the management options. Similar number of homesteads from each category were selected randomly for the study. Household heads or his representative of the selected homesteads was interviewed using a structured questionnaire to collect information relating to the objectives of the study.

At first the age and total area of each homestead were noted; then all components such as plants including crops, trees, shrubs, and herbs; poultry; livestock; and fish in each homestead were identified in charactering the homestead production system. Tree and shrub species with a minimum diameter at breast height (DBH) of 5 cm were counted to determine the population of each tree species for calculating diversity index.

All the identified and counted tree and shrub species were used to assess species richness according to the Shannon index. The Shannon diversity index (*H*) is commonly used for assessing species diversity, which was measured using the formula as described by Abebe et al. (2013) and Magurran (1988):

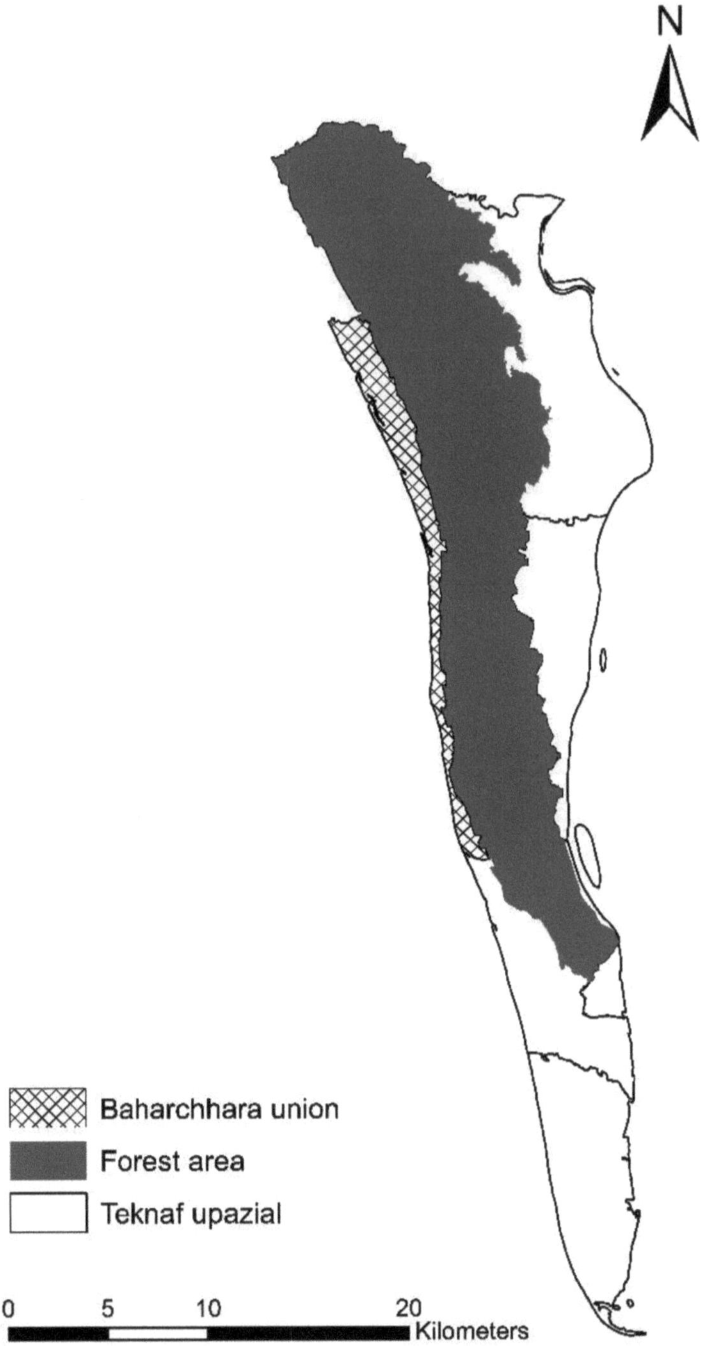

Fig. 12.1 Map showing Baharchara union where the study was conducted

$$H = -\sum PilnPi,$$

where Pi denotes the proportion of individuals composed of species i. This index is also used to describe the abundance and density of species within homesteads.

To calculate the total income of a household, incomes from different sectors such as agriculture, fishing, forestry, services, labor, business, remittances, the homestead, and other sources were totaled. The income derived from the homestead was calculated separately to determine its contribution to the total household income. Homestead income comprised the values of herbs, fruits, vegetables, livestock, poultry, and fish from the home garden that were either consumed or sold. The values of the products obtained from the homestead were calculated based on their market prices during the preceding year.

Farmers plant trees and cultivate crops in different locations of the homestead according to their own preferences and available space. They rely on culture and tradition rather than science when selecting trees and crops. In homesteads, herbs, shrubs, creepers, and trees occur within a multistoried structure in which there is a vertical and horizontal arrangement of species. In this study, the horizontal structure was characterized based on the locations of the species, whereas the vertical structure was characterized based on the heights of different species.

12.3 Results and Discussion

12.3.1 Demographic Characteristics

The demographic characters such as ages of household heads, education levels, family size, and number of earning members are shown in Fig. 12.2.

Household heads belonging to the landless farmer category were generally younger (their average age was 39.6 years) than those belonging to the other categories. Among the other categories, household heads of the medium-sized farm category were relatively old (71.4 years), followed by the large (55.0 years), marginal (51.2 years), and small (48.9 years) farm categories. Young respondents were newly married and were separated from joint family, while some were migrated from other areas.

Education plays a vital role to increase people's awareness in different ways. The educational levels in the study area were found to be extremely low. In general, old respondents (over 50 years) had no formal education regardless of farm categories. Few household heads of landless and small farm categories had a primary level education, whereas no formal education was found for other farm categories (Fig. 12.2). This finding may be attributed to the scarcity of educational institutions in the area. Ullah et al. (2014) found that most of the population on the Teknaf peninsula is illiterate, with only 8% having attained a secondary education. Foysal et al. (2013) reported that educational levels of farmers play a vital role in tree plantation and management.

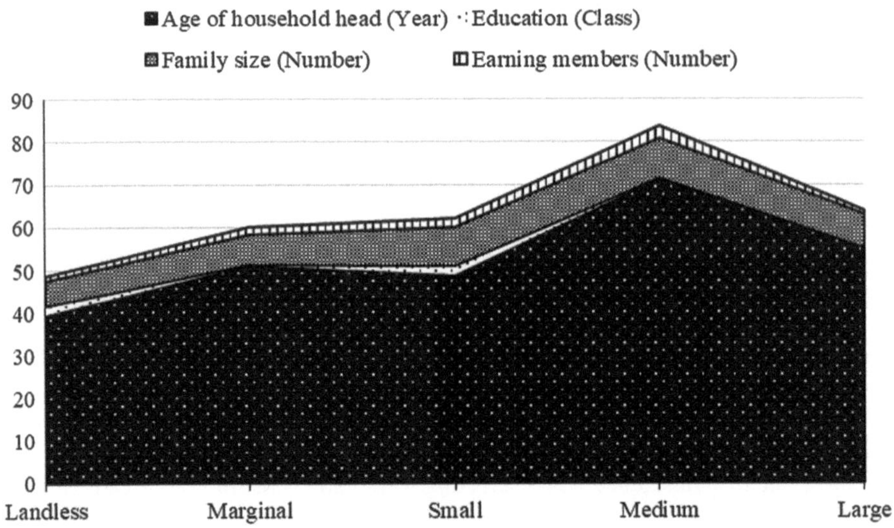

Fig. 12.2 Distribution of the respondents of different farm categories based on age, education, family size, and earning members of the household. Education level indicates the schooling year

The average family size in the study area is about 8, which was quite larger than the national average of 4.5 (BBS 2011). Among households belonging to the different farm categories, the largest family size was found for medium-sized farms, followed by small, large, and marginal farms and landless households.

In addition to household heads, other earning members of a household also play a key role in income generation. Numbers of earning members per household were highest, on average, within the medium-sized farm category (3), followed by the small (2.18) and marginal farm categories (1.83) and landless households (1.21). Notably, numbers of earning members were lowest in households in the large farm category (1.0).

12.3.2 *Time Span Following Settlement and Homestead Areas*

As Fig. 12.3 shows, the time span following the settlement of homesteads varied according to farm categories in the study area. It was less for the landless category, becoming longer with an increase in the size of homesteads. The average time of homestead settlement for landless, marginal, small, medium-sized, and large farm categories were 17.2, 31.5, 31.5, 37.6, and 50.0 years, respectively. This indicates that landless households live in comparatively newer houses than households belonging to other categories. By contrast, houses with large homesteads are old. The average area of homestead in the study area was 0.28 ha. The homestead area was well in correspondence with farm category, where 0.11, 0.18, 0.23, 0.36, and 0.51 ha, respectively, belonged to the landless, marginal, small, medium, and large farm categories.

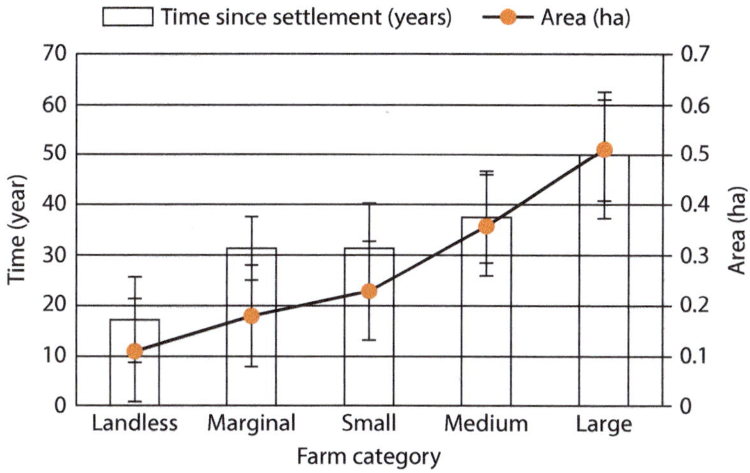

Fig. 12.3 Distribution of the households of different categories based on the length of settlement time and homestead areas. *Vertical bars* indicate standard deviation

12.3.3 Components Found in the Homestead

In a typical homestead of Bangladesh, various components such as vegetable, fruit, timber, medicinal plants, fish, livestock, and poultry are found. Among the tree species, fruit species were commonly found in all the studied homesteads of large, medium, and small farm categories, while it was present in 88.9 and 72.2% homesteads for the marginal and landless categories, respectively. On the other hand, large, medium, and small farms had livestock and poultry. A wide range of vegetables are grown in the homesteads, which were mostly cultivated in the medium (61.1%), small (61.1%), and marginal (50.0%) farm categories. Regardless of farm categories, few of the households in the study area had planted timber species in their homesteads. Medicinal plant species were found in approximately 10% of the homesteads of households with large and medium-sized farms. Fishponds were not found in any of the homesteads (Fig. 12.4).

12.3.4 Seasonal Crops Grown in Homesteads

A wide range of seasonal crops that include vegetables and spices are generally cultivated in the homesteads to meet the daily need of the household, and additional parts are sold in the market for income generation, in some cases. Interestingly, the homestead production system is managed by family members. Among the seasonal vegetables, papaya (*Carica papaya*), aroid (*Alocasia indica*), and eggplant (*Solanum melongena*) were grown throughout the year. Whereas, in summer season bitter gourd (*Momordica charantia*), okra (*Abelmoschus esculentus*), ash gourd (*Benincasa*

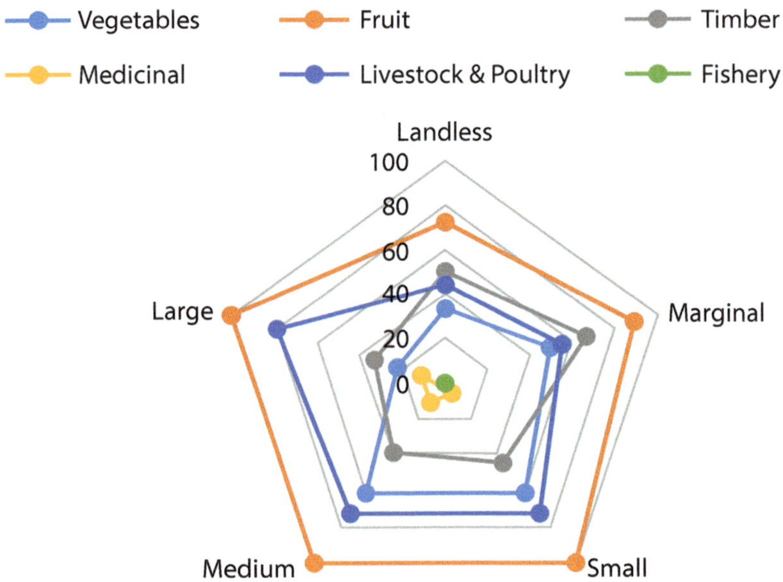

Fig. 12.4 Components of the homestead production systems of different farm categories

hispida), ribbed gourd (*Luffa acutangula*), taro (*Colocasia esculenta*), yard-long beans (*Vigna unguiculata*), bottle gourd (*Lagenaria vulgaris*), cucumber (*Cucumis sativus*), bush beans (*Phaseolus vulgaris*), and red amaranth (*Amaranthus* sp.) were grown. Although many winter vegetables are grown in other parts of the country, it was found few in the studied homesteads as water is scarce in the study area particularly during winter season. However, tomato (*Solanum lycopersicum*), potato (*Solanum tuberosum*), and country bean (*Phaseolus vulgaris*) were some common winter vegetables grown in homesteads in Teknaf (Fig. 12.5). Among the spices, chili (*Capsicum annuum*), turmeric (*Curcuma longa*), ginger (*Zingiber officinale*), and coriander (*Coriandrum sativum*) were commonly grown in the homesteads.

12.3.5 Plant Species Arrangement

A lot of plant species are grown in homestead with different natures, which are arranged mainly as a horizontal (spatially distributed) and as a vertical (stratified). Diversified plant species comprising fruit trees, timber species, crops, vegetables, herbs, shrubs, medicinal plants, and bushes are found at different locations of the homesteads. The horizontal arrangement of the species was separated in several locations such as the backyard, front yard, boundaries, corners, and other areas, and it was found varying species composition in different locations. Among the locations, back- and front yards had more trees in terms of number of individual species than other locations. On the other hand, single species with high density was found

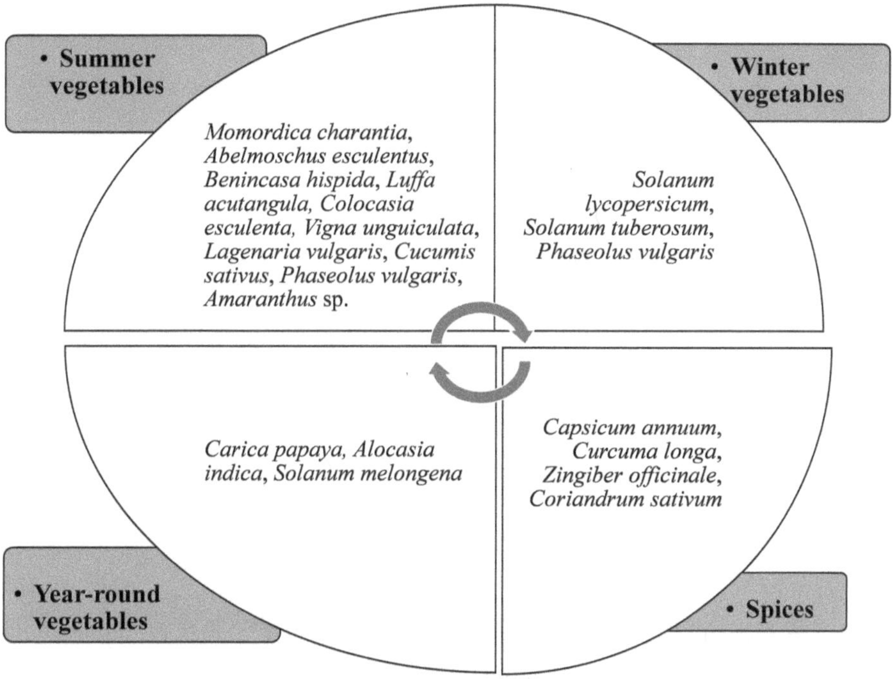

Fig. 12.5 Season-wise vegetables and spices grown in the homesteads

at the boundaries (*Areca catechu* and *Cocos nucifera*) and corners (*Bambusa balcooa*) of the homesteads (Fig. 12.6). At the front yards where there was sufficient light, various types of vegetables and herbs are grown. In other areas such as marshy land, mainly different aroids (*Alocasia indica*) were found.

To determine the vertical structure of homestead vegetation, we distinguished different strata: less than 1 m, 1–4 m, 5–7 m, 8–10 m, and greater than 10 m (Fig. 12.7). Both the horizontal and vertical vegetation structures revealed that homesteads on the Teknaf peninsula were diversified with regard to plant species composition. Mainly different fruit and timber species such as coconut (*Cocos nucifera*), garjan (*Dipterocarpus* sp.), mango (*Mangifera indica*), jackfruit (*Artocarpus heterophyllus*), and bamboo (*Bambusa balcooa*) were found at the top layer. Small trees, shrubs, and vegetables comprised the lower layer. However, the predominant tree species, betel nut (*Areca catechu*), was found in almost all of the homesteads within different locations and strata.

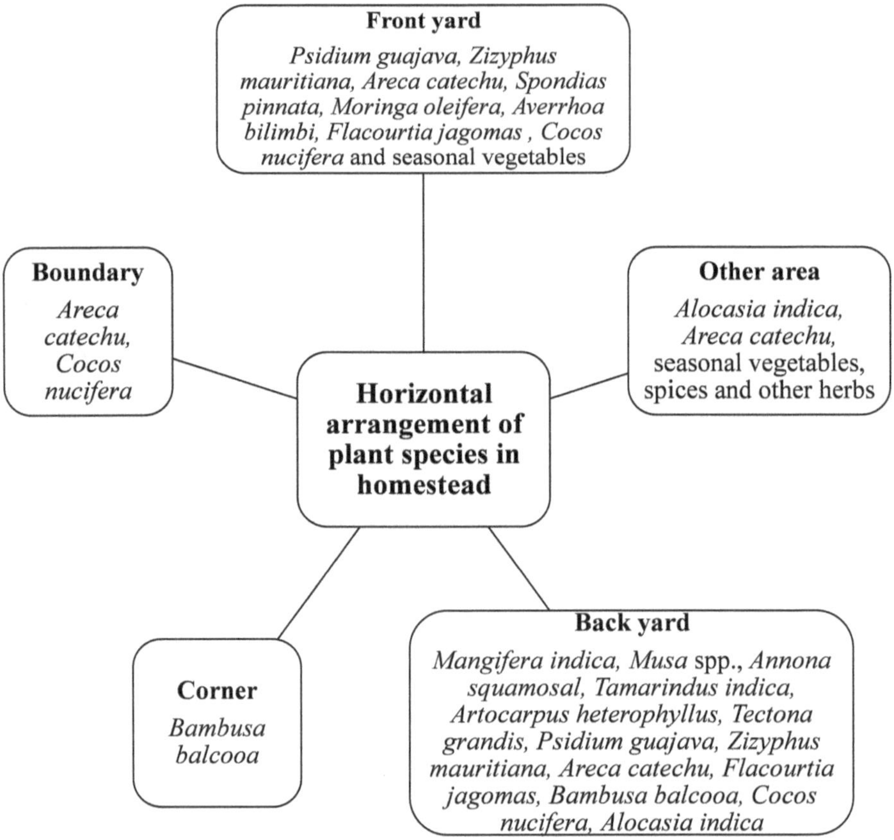

Fig. 12.6 Horizontal arrangement of different plant species in a homestead

12.3.6 Tree Diversity

In the studied homesteads, for all categories, about 160 tree, shrub, and herb species were recorded. The number of tree species varied remarkably in homesteads of different categories, where over 85% homesteads have about 17 tree species. The average number of trees per homestead varied widely across the farm categories (Fig. 12.8). Homesteads in the medium farm category had the most tree species, followed by those in the large farm category. The lowest number of individual trees was found in homesteads belonging to landless farmers. Homesteads of households in the large farm category contained the highest number of trees (363.7), which was significantly higher than those found in homesteads belonging to households in the other farm categories. Landless farm category has significantly lower number of trees in the homesteads compared to those found in homesteads of households in the other farm categories.

Shannon's diversity index showed that homesteads of households in the large farm category had the highest species diversity, followed by those of households in

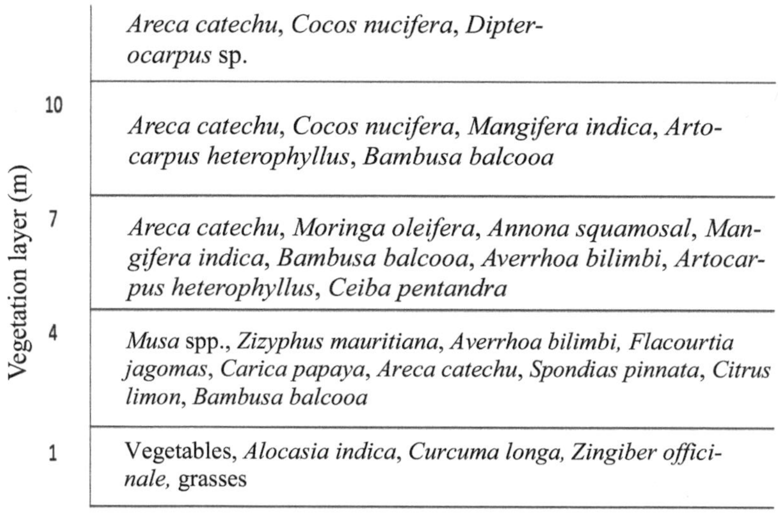

Fig. 12.7 Vertical arrangement of different plants grown in homestead

the medium, small, and marginal farm categories and lastly those of landless households (Table 12.1). In the studied homesteads, betel nut (*Areca catechu*) tree was the most dominant species, irrespective of farm categories, followed by mango (*Mangifera indica*), jackfruit (*Artocarpus heterophyllus*), and coconut (*Cocos nucifera*) tree species.

12.3.7 Annual Homestead Incomes in Relation to Total Annual Household Incomes

Total annual household incomes were lowest for landless farmers, gradually increasing for households in the larger landholding groups up to the large farm category. Total annual incomes were highest for households with homesteads in the large farm category, followed by those with homesteads in the medium-sized farm category. Annual total incomes of landless farmers with homesteads were almost half than those of households with homesteads in the large farm category. Although a similar trend was found for annual homestead incomes, these incomes were appreciably higher for households in the large farm category compared with those of households in the other categories. However, annual homestead incomes were not high, and the highest and lowest incomes were found among households in the large farm and landless farmer categories, respectively. As shown in Fig. 12.9, the contribution of homestead incomes to annual total household incomes was highest (38.1%) for those in the large farm category, followed by those in the marginal (26.2%) and medium-sized (25.7%) farm categories.

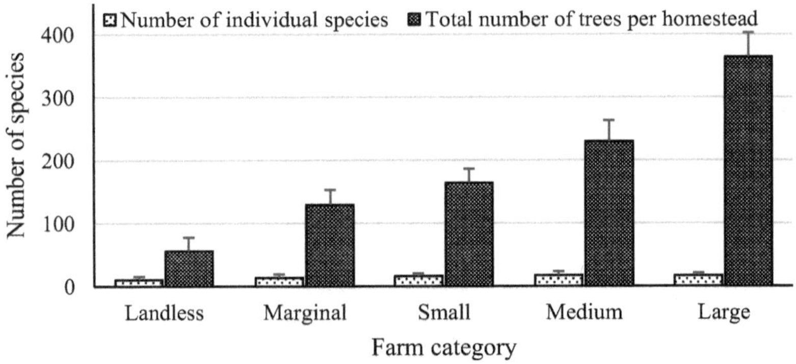

Fig. 12.8 Mean number of individual tree species per homestead and number of tree species in homesteads belonging to different farm categories. Trees of >5 cm DBH were taken in the calculation. *Vertical bars* indicate standard deviations

Table 12.1 Shannon–Weaver diversity index of tree species (>5 cm DBH) in the studied homesteads

Tree	Landless	Marginal	Small	Medium	Large	Average
Spondias pinnata	0.05	–	0.10	0.04	0.05	0.06
Annona squamosa	0.06	0.04	0.06	–	0.03	0.06
Averrhoa bilimbi	0.03	–	–	0.06	0.05	0.05
Nauclea latifolia	–	0.04	0.04	0.03	0.04	0.04
Ziziphus mauritiana	–	0.03	0.12	0.08	0.05	0.07
Cocos nucifera	0.10	0.06	–	0.08	0.29	0.13
Gmelina arborea	0.03	0.03	0.03	–	0.07	0.04
Dipterocarpus sp.	0.13	0.03	0.07	0.09	0.08	0.08
Psidium guajava	0.07	0.04	0.11	0.15	0.05	0.08
Artocarpus heterophyllus	0.10	0.20	0.18	0.17	0.20	0.17
Mangifera indica	0.21	0.30	0.25	0.28	0.27	0.26
Flacourtia jangomas	0.03	0.10	–	0.09	0.03	0.06
Carica papaya	0.05	0.04	0.04	0.04	0.07	0.05
Moringa oleifera	0.03	0.08	0.07	0.08	0.06	0.06
Dalbergia sissoo	0.05	0.03	0.03	–	0.02	0.03
Areca catechu	0.30	0.32	0.33	0.35	0.37	0.34
Ceiba pentandra	0.12	0.09	0.09	0.09	–	0.10

12.4 Conclusions

Homesteads are dynamic production systems that are prevalent in Bangladesh. The time span following establishment of the homesteads examined in this study varied from 17.2 to 50.0 years. The average homestead area was 0.28 ha, which was relatively larger than homestead areas in other parts of Bangladesh. Among the components, almost all homesteads contained various plants such as fruit, timber, and

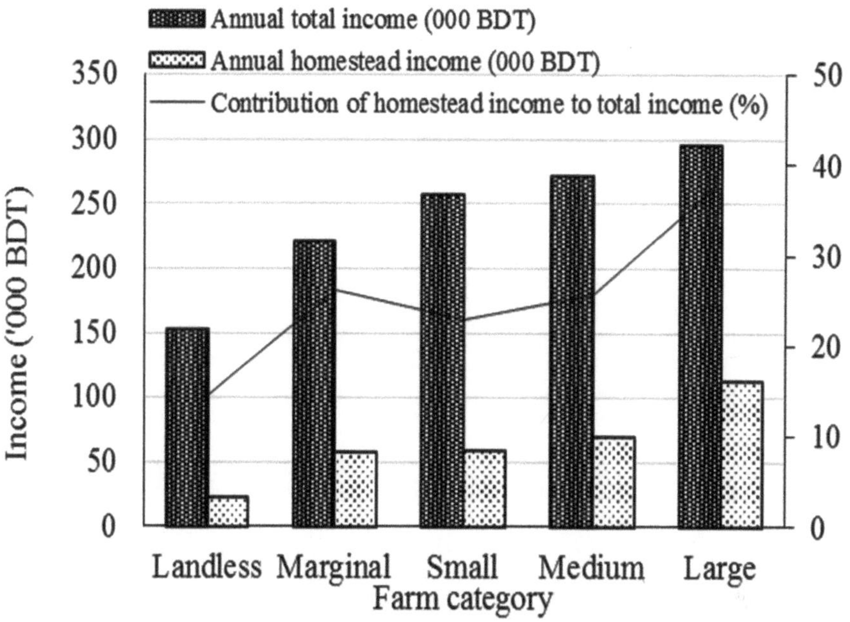

Fig. 12.9 Household total income, homestead income, and percent contribution of homestead income to the total annual income by farm category. *BDT* Bangladeshi Taka

vegetable species; livestock; and poultry although their distribution and density varied between farm categories. The investigation of species arrangement in terms of horizontal and vertical arrangements showed no specific pattern of species distribution. Although a lot of tree species were identified in the homesteads, betel nut tree was the most dominant, regardless of farm categories, which contributed significantly to annual income of the households. We suggest that homestead incomes could be increased by designing an efficient homestead production system. Moreover, fuelwood, timber, fruits, and medicines can be obtained from homesteads through the appropriate selection of compatible species.

References

Abebe T, Sterck FJ, Wiersum KF (2013) Diversity, composition and density of trees and shrubs in agroforestry homegardens in southern Ethiopia. Agrofor Syst 87:1283–1293

Abedin MZ, Quddus MA (1988) Homestead trees and household fuel uses in and around the farming systems research site, Jessore. In: Abedin et al. (eds) Homestead plantation and agroforestry in Bangladesh. Proceedings of a National Workshop held on 17–19 July, 1988 at Joydebpur, Gazipur, Bangladesh. BARI/RWEDP/WINROCK, pp 106–119

Ali MA, Momin MA, Alam MR, Mollah MSH, Ivy FI, Hossain MA (2006) Goyespur vegetable production model (a Bengali booklet). On-Farm Research Division, Bangladesh Agricultural Research Institute, Pabna

BBS (2011) Preliminary report of census. Bangladesh Bureau of Statistics, Ministry of Planning, Government of the People's Republic of Bangladesh, Dhaka

Chowdhury MK, Satter MA (1992) Homestead and crop land agroforestry practices in the high Ganges floodplain. Research Report Series to BARC Winrock Intl, Dhaka, pp 23–55

Fernandes ECM, Nair PKR (1990) An evaluation of the structure and function of tropical homegarden. In: Kathleen L, Mark B (eds) Tropical Homegarden. United Nations University Press, Tokyo, pp 105–114

Foysal MA, Hossain ML, Rubauyat A, Hasan MB (2013) Economics of homestead forestry and their management activities at Fatickchari upazila of Chittagong district, Bangladesh. Agric For Fish 2(4):161–176

Halladay P, Gilmour DA (1995) Conserving bio-diversity outside protected area. The role of traditional agro-ecosystem. IUCN Publication, Gland, pp 172–175

Iannotti, L, Cunningham K, Ruel M (2009) Improving diet quality and micronutrient nutrition: homestead food production in Bangladesh. IFPRI Discussion Paper. International Food Policy Research Institute, Washington DC

Kabir ME, Webb EL (2008) Can homegardens conserve biodiversity in Bangladesh? Biotropica 40(1):95–103

Landauer K, Brazil M (eds) (1990) Tropical home gardens. Selected papers form an international workshop at the Institute of Ecology, Padjadjaran University, Indonesia, December 1985. United Nations University Press, Tokyo

Magurran AE (1988) Ecological diversity and its measurement. Croom Helm, London

Marsh R (1996) Household gardening and food security: a critical review of the literature. Nutrition Programmes Service, Food and Nutrition Division, FAO, Rome

Miah MG, Hossain MK (2001) Study of the indigenous knowledge on the homestead forestry of Narsingdi region, Bangladesh. South Asian Anthropol 1(2):129–135

Millat-E-Mustafa HJB, Teklehaimanot Z (1996) Structure and floristics of Bangladesh homegardens. Agrofor Syst 33(3):263–280

Rahman MA, Tani M, Asahiro K, Ullah SMA (2016) Spiece composition, diversity and productivity of homesteads in Southeastern Bangladesh. Small-scale Forestry, DOI 10.1007/s11842-016-9356-8

Roy I, Siddiqui AR, Huda ATMA (1996) Village and farm forestry Programme: an analysis of trend to plant for VFFP phases (1997–2001): an input for evaluation cum planning Mission Village and farm forestry project. Swiss Agency for Development and Cooperation, Dhaka

Ullah SMA, Tsuruta H, Tani M, Rahman MA (2014) A comparative study of socioeconomic characteristics between two villages in the Teknaf Peninsula. In: Proceedings of the International Conference on Environmental Aspects of Bangladesh (ICEAB 2014). University of Dhaka, Dhaka

Zaman S, Siddiquee SU, Katoh M (2010) Structure and diversity of Homegarden agroforestry in Thakurgaon District, Bangladesh. Open For Sci J 3:38–44

Chapter 13
Land Use and Composition of Homestead Forests

K. Asahiro

Abstract The Teknaf peninsula is experiencing a decline in its natural resources, which are being used to support the local population and ecology. Forest destruction is a particularly serious problem in the area. Every household has a homestead, which is an important indicator of an improved local lifestyle. This chapter examines the land use and composition of homesteads in the village of Marishbunia, which is located in the western coast of the Teknaf peninsula. As a first step, it presents a mapping of the general land use and vegetation in this village based on a survey covering 46 short transects lines. High-resolution satellite images of the area were also analyzed. The results of our mapping of general land use indicated the following sequence of vegetation from the coast extending to the hilltop: sand, windbreak, farmland, the main road and homestead gardens, betel leaf fields, social forests, sungrass fields, and mantle communities around hillsides and on hilltops. The destruction of national forests is thought to be increasing because of wood and grass harvesting. On the other hand, homestead gardens provide valuable vegetation, including tall trees in this region. To develop an understanding of the homestead lifestyle and vegetative environment, seven homesteads were selected for a survey conducted on the social dimensions, vegetation, and number of trees with a diameter at breast height (DBH) of more than 5 cm. A total of 193 species were identified, of which 44 occurred in more than 50% of the surveyed homesteads. Because betel nut cultivation seems to be on the rise, there should be greater emphasis on diversifying tree plantations.

Keywords Landscape • Vegetation composition • Hill forests • Betel leaf farming • Homestead garden

K. Asahiro (✉)
Department of Environmental Design, Faculty of Design, Kyushu University, Fukuoka, Japan
e-mail: asahiro@design.kyushu-u.ac.jp

M. Tani, M.A. Rahman (eds.), *Deforestation in the Teknaf Peninsula of Bangladesh*, DOI 10.1007/978-981-10-5475-4_13

13.1 Introduction

Most of the households in the Teknaf peninsula get forest products easily including fuelwood, timber, fodder, and various other products, as well as non-timber forest products. Therefore, people in Teknaf do not take much care about their homesteads. The over dependency of the local people on forests creates tremendous pressure on forest resources. To alleviate forest degradation, two different sets of strategic priorities have been described by Iftekhar (2006). One is maintaining the environmental and ecological services of the forests through multi-stakeholder-based holistic management. The other is meeting forest product demand from homesteads and private forests through massive forest extension and social forestry activities. It is also reported the importance of homesteads for biodiversity conservation. A total of 419 species have been identified in 402 homesteads in southwestern Bangladesh, but 60% of all tree and shrub species had 50 or fewer individuals. Thus, a serious effort must be made to increase the populations of most species (Kabir and Webb 2008).

The Teknaf peninsula is an ecologically critical area (ECA) of Bangladesh (CWBMP 2011). In the Teknaf ECA, 249 species of plants belonging to 84 families have been identified in 65 places, which is evidence of a decreasing trend of homestead plant biodiversity (Mannan 2006). The Teknaf Wildlife Sanctuary (TWS) was explored from 2010 to 2011 by Muhammed et al. (2008) to assess angiosperm diversity using traditional taxonomic techniques. The assessment recorded 535 angiosperm species belonging to 370 genera and 103 families (Uddin et al. 2012). Although traditional agroforestry is being practiced in Teknaf, its potential benefits have not been tapped because of neglect and improper planning and management. In these situations, homestead forests should be considered in terms of production, biodiversity conservation, and quality of life maintenance.

This chapter reports a case study of homesteads in Marishbunia (MB) village on the west coast of the Teknaf peninsula to describe the following points:

1. The general landscape composition of the homesteads
2. The general distribution of trees and other vegetations in homestead gardens
3. A characterization of the homesteads based on their settlement history and tree species

13.2 Methodology

13.2.1 General Landscape Composition

The west coast of the Teknaf peninsula includes three unions and six villages. MB is one of the villages in the Baharchara union, and it was selected for this study because it spans the short distance from the coast to the hilltop and comprises a wide range of land uses and vegetation (Fig. 13.1).

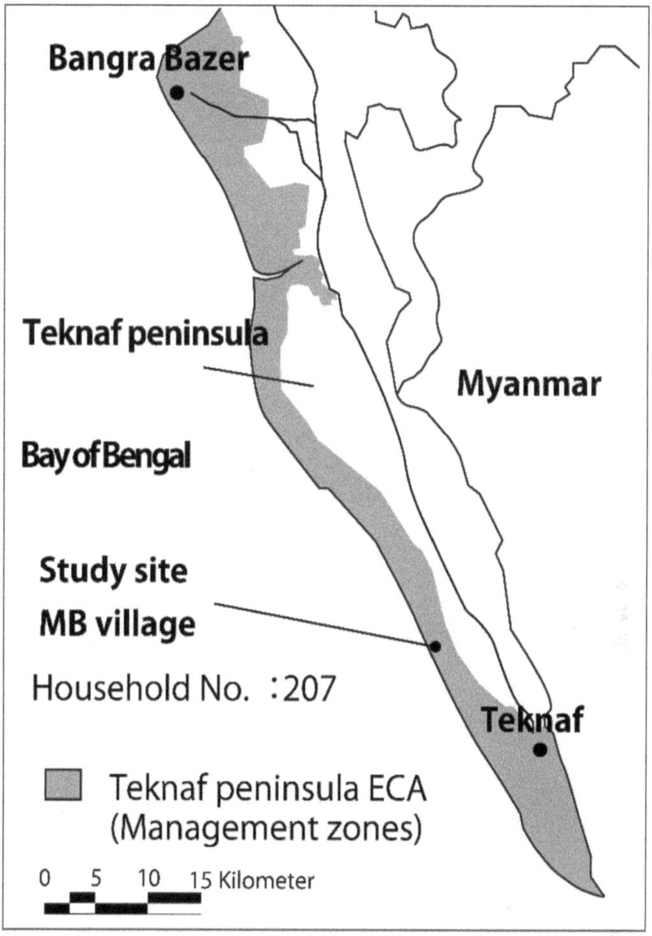

Fig. 13.1 Position of Marishbunia (MB) village

Investigations were conducted in March 2010. The methodology used in this study involved understanding the general land use and vegetation distribution using 46 short transect lines that were approximately 20 m in length. In each site, the lengths and positions of the transect lines, dominant species, and the height and cover of each vegetation layer were recorded. Pictures of vegetation were also taken.

After the survey, 11 types of land use and vegetation were categorized based on the ground research data. Then, an interpretation of high-resolution satellite images was conducted. The vegetation data were used in the following equations to determine the canopy cover (13.1) and species composition (13.2).

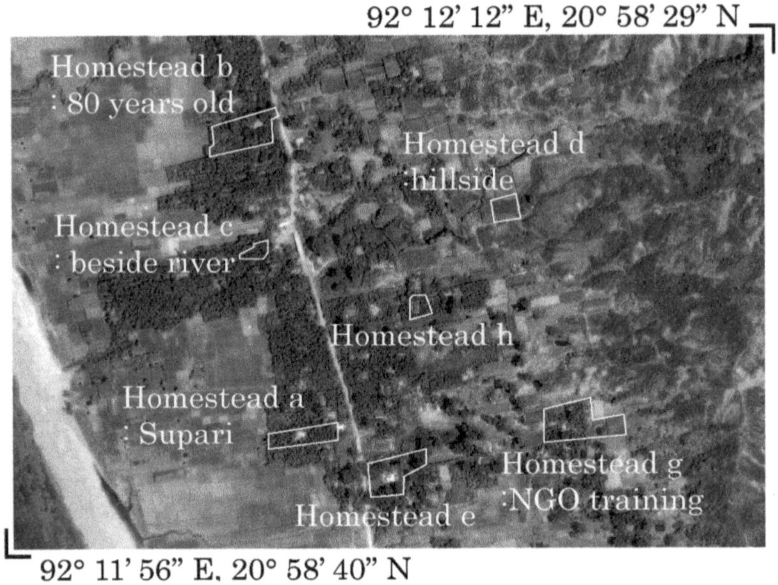

Fig. 13.2 Selected homestead gardens (outlined in *white*)

$$\text{cover}(\%) = \frac{\text{Total intercept length of species}}{\text{Total transect length}} \times 100 \quad (13.1)$$

$$\text{Species compostion}(\%) = \frac{\text{No. of individuals of a species}}{\text{Total no. of individuals of all spp.}} \times 100 \quad (13.2)$$

13.2.2 General Distribution of Trees and Other Vegetations

To understand the composition of trees and their structure, seven homesteads were selected, primarily based on their location. Homesteads a, b, and c are located on the western side of the main road (and are considered to constitute a comparatively old settlement), while the other four homesteads (d, e, f, and g) are located on the eastern side of the main road (a new area that is a part of the national forest) (Fig. 13.2). Additionally, homestead c is located near a stream, while homestead d is located in a hilly region, and homestead g is owned by a person who received tree planting training from a non-governmental organization.

To determine the boundaries of the homesteads, the head of household/resident was asked to walk around the boundary. The positions of the fence, house, toilet, and entrance were determined by simple measurements with a compass and measuring tape. To determine tree growth and composition, trees, excluding betel nut (*Areca catechu* L.) trees, with a diameter at breast height (DBH) greater than 5 cm were selected.

13.2.3 Characterization of the Homesteads Based on Their History and Tree Census

To characterize the homesteads, the following three investigations were conducted in September 2010:

1. *Survey for producing maps*: Homestead boundaries, houses, forests, crop fields, and tree positions were recorded.
2. *Interview survey*: Farming areas and activities were recorded by interviews.
3. *Vegetation and tree census*: Tree layer areas were surveyed using the Braun-Blanquet method. Species and DBH measurements were conducted as described above. Betel nut trees were counted, and their density within a 5-m radius area was recorded.

13.3 Results and Discussion

13.3.1 General Landscape Composition

The short transect plots (determined by Global Positioning System measurements at the edge of each line), land use, and vegetation type map are depicted in Fig. 13.3. The map shows the general land use and vegetation sequence from the coast to the hilltop, including the beach, tree plantation (windbreak), farmland, main road, and homestead gardens on both sides of the road. There are also *pan boroj*, which are used to grow betel leaf, between the main road and the hill, a social forestry plot on the hillside, and sungrass (*Imperata cylindrica*) fields and scrub on the hillside and hilltop areas.

The percentage of vegetation cover of the dominant species for each land use, vegetation type, number of transect lines, number of species, and species composition (%) are shown in Table 13.1. The general land uses and vegetation types are listed by category, which are described below (Fig. 13.4).

Windbreak Jhau (*Casuarina equisetifolia*) was planted as a windbreak, and it covered 50% of the coastline. However, its continuity along the shore is broken, and the width of the windbreak belt is only approximately 10 m. In the coastal area, incidences of windbreak clearing have been reported. In addition to jhau, akand (*Calotropis gigantea*), and chagol kuri (*Ipomoea pes-caprae*) are also distributed along the seashore.

Farmland Farmland is spread along the windbreaks and homestead gardens. In the rainy season, rice is grown in wet paddy fields, while irrigation water from the stream and wells are used to water crops, such as beans and vegetables, in the dry season.

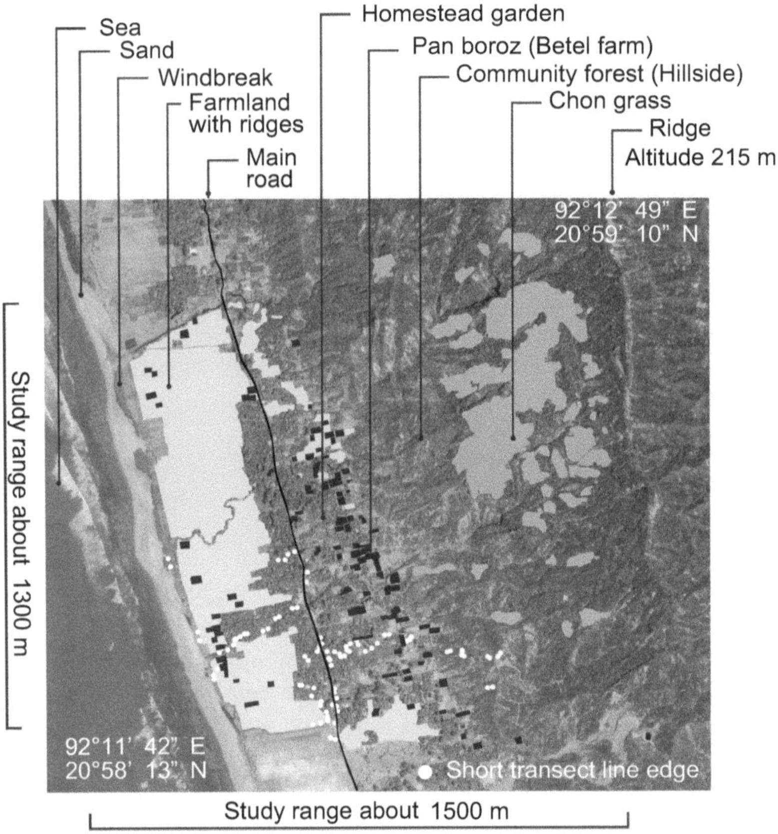

Fig. 13.3 Land use and vegetation type map generated by satellite image interpretation

Homestead Gardens Approximately 207 households with thatched roof houses and homestead gardens occupy both sides of the main road. Betel nut trees were planted in and around the houses and covered 33% of the homestead area, making betel nut the most dominant species. Along the roadside, mango (23.8% of the homestead area) and betel nut (16.6% of the homestead area) were dominant, along with other fruit and crop trees, as well as shrub species. Another tree that was prominent in the village was garjan (*Dipterocarpus alatus*, accounting for 8.8% of the area along the roadside), and these trees were nearly 26 m tall. Among the homestead gardens, bamboo (*Bambusa* spp., 18.6%) and various native tree species were observed. These included banana (*Musa sp.*, 4.3% of the homestead area), fig (*Ficus racemosa,* 1.4% of the homestead area), sheora (*Streblus asper*, 1.4% of the homestead area), and khoi (*Bridelia tomentosa*, 1.4% of the homestead area). These species likely propagated from seeds brought down by the stream from the hilltop. In terms of species composition, 48.1% of the species, covering 25 of the 46 transect lines, were found inside the homestead gardens. The actual species composition in the homestead gardens was estimated to be lower than that in the hilly areas.

Table 13.1 Coverage (%) of dominant species for each land use and vegetation type

Layer	Transect line number	Tree height (m)	Bengali name	Scientific name	Windbreak	Seashore vegetation	Farmland with ridges	Homestead garden (roadside)	Homestead garden (inside)	Homestead garden (riverside)	Dug well	Community forest (hillside)	Mantle community (hillside to hilltop)	Chon grass (hillside to hilltop)
Tree layer	13	10	Betel nut	*Areca catechu* L.				16.6	33.0					
	5	12	Bhalku	*Bambusa* sp.			1.3	6.3		18.6				
	5	11	Mango	*Mangifera indica* L.				23.8	4.0					
	5	7	Coconut	*Cocos nucifera*			6.3	2.5	1.3					
	3	26	Garjan	*Dipterocarpus alatus*				8.8	1.0		2.5			
	4	11	Jackfruit	*Artocarpus heterophyllus*				1.6	6.0					
	1	14	Jhau	*Casuarina equisetifolia*	50.0									
	1	6	Banana	*Musa* spp.						4.3				
	1	10	Fig	*Ficus racemosa*						1.4				
	1	10	Sheora,	*Streblus asper*						1.4				
	2	13	Khoi	*Bridelia tomentosa*						1.4				1.5
	1	7	Akashmoni	*Acacia auriculiformis*								90.0		
Shrub layer	5		Dhol kolmi	*Ipomoea fistulosa*			10.0	0.4		0.4	22.5			
	6		Betel nut	*Areca catechu* L.				9.5	2.0		0.8			
	3		Mahogini	*Swietenia mahagoni*				1.3	1.0	0.4				
	1		Akand	*Calotropis gigantea*		35.0								
	2		Bhalku	*Bambusa* sp.			1.3	0.4						
	1		Kul, boroi	*Ziziphus mauritiana*			8.8							

(continued)

Table 13.1 (continued)

Layer	Transect line number	Tree height (m)	Bengali name	Scientific name	Windbreak	Seashore vegetation	Farmland with ridges	Homestead garden (roadside)	Homestead garden (inside)	Homestead garden (riverside)	Dug well	Community forest (hillside)	Mantle community (hillside to hilltop)	Chon grass (hillside to hilltop)
Grass layer	8		Badaiya	*Chrysopogon aciculatus*			35.0		14.0	10.0	17.5			
	3		Chhagol Kuri	*Ipomoea pes-caprae*	10.0	20.0								
	3		Guicha lata	*Calycopteris floribunda*				0.4	7.0				3.3	
	4		Betel nut	*Areca catechu* L.				0.4	1.3	4.3				
	4		Assam lata	*Mikania cordata*					5.3		13.3			
	2		Sungrass	*Imperata Cylindrica*										50.0
	3		Germanlata	*Chromolaena odorata*				0.4	7.0		0.8			
	1		–	*Phyllanthus pendulus*									10.0	
	1		Mashkalai	*Vigna mungo*			8.8							
	1		Bhalku	*Bambusa* sp.			1.3							
	1		Aroid	*Colocasia esculenta*							2.5			
	1		Arjun	*Terminalia arjuna* Bedd.					1.0					
Number of transect lines					1	2	8	8	10	7	4	1	3.0	2
Number of species					2	2	10	8	13	7	7	1	6.0	2
Species composition (%)					7.4	7.4	37.0	29.6	48.1	25.9	25.9	3.7	22.2	7.4

Length of transect line, 20 m; no. of transect lines, 46

Fig. 13.4 Types of land use and vegetation in the study area

Although the transect lines through the hill areas in this study were narrow, only a limited part of the natural hill vegetation was observed. Furthermore, some tall trees of various species were also observed in the homestead gardens.

Pan Boroj In general, betel leaf is the main cash crop in this area, and it requires *pan boroj* structures, which are made of thin timber poles, grass, and other minor forest products, for shade. These construction materials are collected mainly from locally available natural vegetation growing on the hillside. In the dry season, betel leaf is irrigated using water from wells, which are located nearby.

Hillside Most of the hillside areas are used as a social forest plot, and they were planted with akashmoni (*Acacia auriculiformis*, which covered 90% of the plot), with shrubs underneath. Sungrass fields are commonly found along the ridges of the hill, where there are almost no tree. The World Wide Fund for Nature has confirmed that 31 wild elephants inhabit this region. According to local residents, wild elephants feed on fruits and crops from homestead gardens, as food availability is scarce on the hill.

From the analysis of land use and vegetation types, it was noted that the destruction of natural forest is occurring because of wood and grass harvesting in the area. In the homestead gardens, betel nut is intensively planted for fruit harvesting. This area also contains other tall trees, including fruit trees, garjan, and other native tree species that propagate by the stream. The homestead gardens, which contain a wide range of plant species, are considered to be valuable for maintaining plants in this region.

13.3.2 General Distribution of Trees and Other Vegetations

The distribution of trees and other vegetation of the homesteads is shown in Fig. 13.5. The average size of the seven homesteads is 1854 m^2, with a range of 480 to 4448 m^2. Every homestead has a boundary comprising fences made of bamboo and/or betel nut leaves. There are also shelters and a central area with hard soil that formed because of activities that are performed in front of the houses. In general, betel nut and fruit trees surround the houses. Some seedlings of fruit species are found, as they grow from seeds that are thrown away after eating fruit. Mango, jackfruit, bamboo, bilimbi, gora neem, and coconut are some common fruit species in the homesteads. Larger homesteads have gardening plots for horticulture and wells (homestead g). Agroforestry systems, with papaya, guava, mango, drumstick, and other fruit species, are found in some homesteads that were established after 2005 (homesteads a, c, d, and g). Some seasonal vegetables and spices are grown in and around the homesteads. Creeping-type vegetables are also grown on trellises (homesteads d and g).

13.3.3 Characterization of the Homesteads Based on Their History and Tree Census

The time of initial settlement, occupation of the householders, and area of the households are shown in Table 13.2. Homesteads a, b, and c were established prior to the 1970s. After the 1980s, homesteads d, e, f, and g encroached into the reserve forest to establish new houses on the west side of the main road. Homesteads a, b, c, e, and f were built recently. Homesteads d and g were moved from other places. Farming was the most common occupation of the householders. However, betel leaf

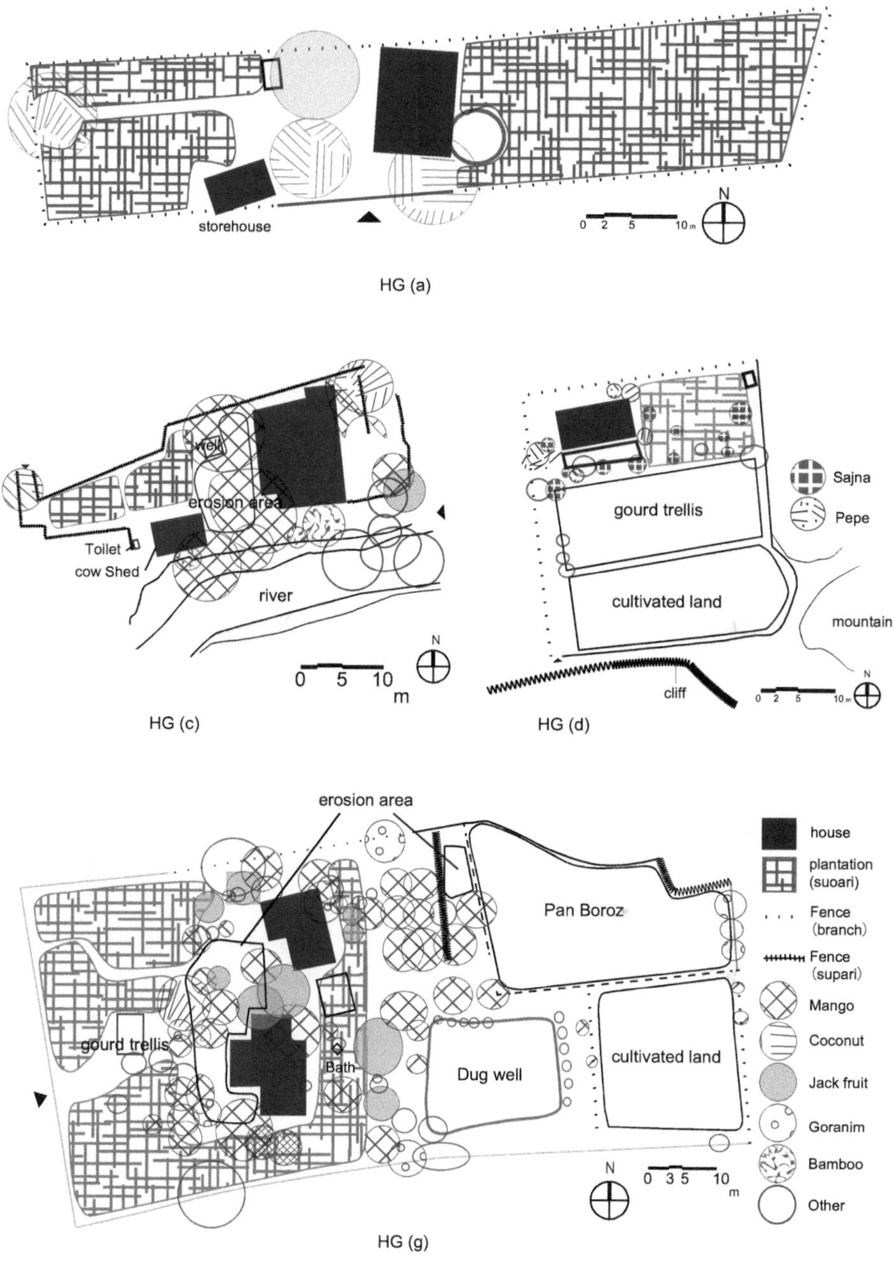

Fig. 13.5 Distribution of trees and other vegetation of the homesteads

Table 13.2 Year of initial settlement, occupation of the householders, and areas of the homesteads and farmland

	Year	No. and (%) of homesteads *	a	b	C	d	e	f	g
History	< 1950s	14 (7)		■	■				
	1950s	1 (0)							
	1960s	5 (2)							
	1970s	11 (5)	■						
	1980s	35 (17)							
	1990s	67 (33)						■	○
	2000s	64 (31)				○	■		
	2010s	7 (3)							
	Unknown	3							
	Total	207							
Occupation	Farming (PB): *Pan* farming)		○	○	○PB	○PB	○	○PB	○
	Side business		○					○	
	Migrant worker						○		
Area	Homestead (m^2)		1452	3024	552	952	2070	480	4448
	Farmland (m^2)		5320	7980	1882	1330	3990	2660	4655

■, new establishment; ○, moved from other places; *, number of homesteads in MB village

cultivation, side businesses, and migrant work were other occupations of the households. In general, the areas of the homesteads were smaller than those of the farmland. The mean farm and homestead areas were 3974 and 1854 m^2, respectively, and farm size varied from 1330 to 7980 m^2.

Figure 13.6 shows the results of the tree census survey. There was a strong relationship between the DBH of the tree species and the age of the homesteads. The highest DBH was noted in homesteads b and c, which were settled prior to the 1950s. Both homesteads consist of three Am trees that are greater than 40 cm DBH, and other large fruit species, such as jackfruit and coconut, were also found. Homestead a, which was settled in the 1970s, owns three coconut trees that are greater than 30 cm DBH. Homesteads d, e, f, and g mainly have trees that are less than 15 cm DBH. Some small tree species were recorded in these homesteads, as they were established recently. Homestead d had papaya (*Carica papaya*) and drumstick (*Moringa oleifera*) trees that can produce economically viable yields within a few years.

Figure 13.7 shows the number of trees, the number of betel nut trees, and the homestead areas. Betel nut was the most dominant tree species for all of the homesteads. The proportion of betel nut trees was highest for homestead a (98.4%), followed by homestead g (71.2%). Trees in homestead g are still young, but many kinds of fruit trees were planted. The head of household of homestead g received

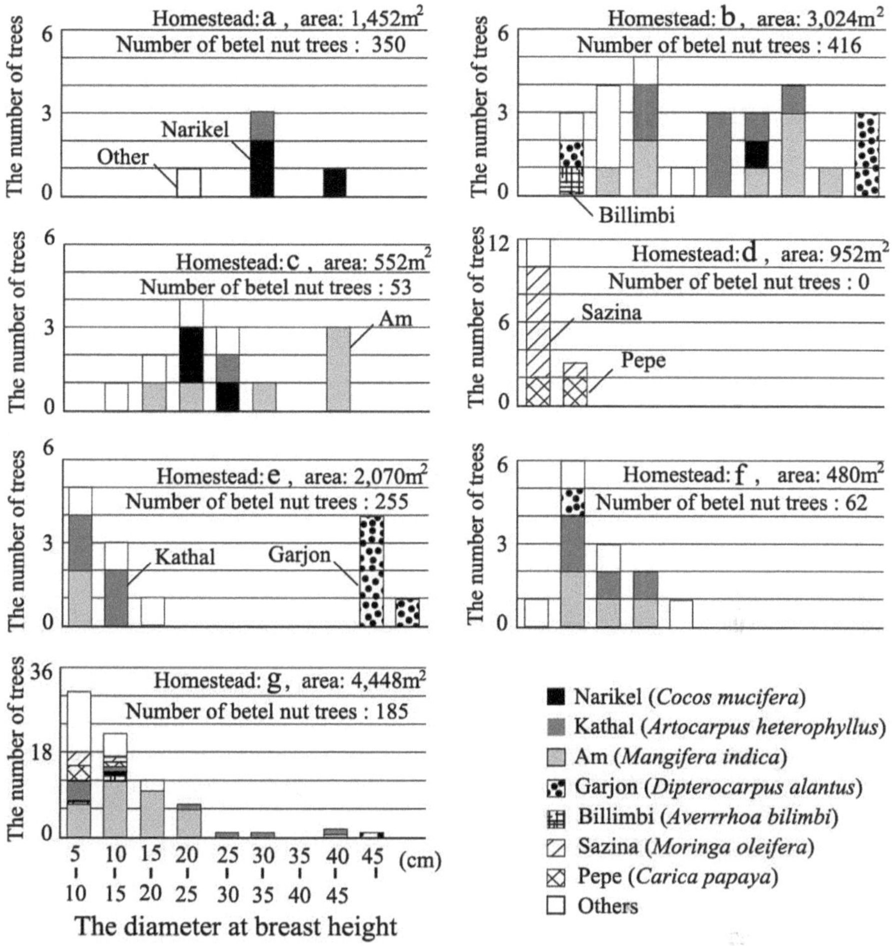

Fig. 13.6 Number of trees (by DBH) of each homestead

tree planting training from the Nature Conservation Management, which may have influenced him to plant many trees on his homestead. This observation is also supported by the findings reported by Rahman et al. (2006), who showed that training was related to tree diversity, land size, and education.

13.4 Conclusions

The age of the homesteads varied from 14 to 65 years. Farming was the main occupation, and the income from agriculture was high for large farms. Other occupations included side businesses and migrant work.

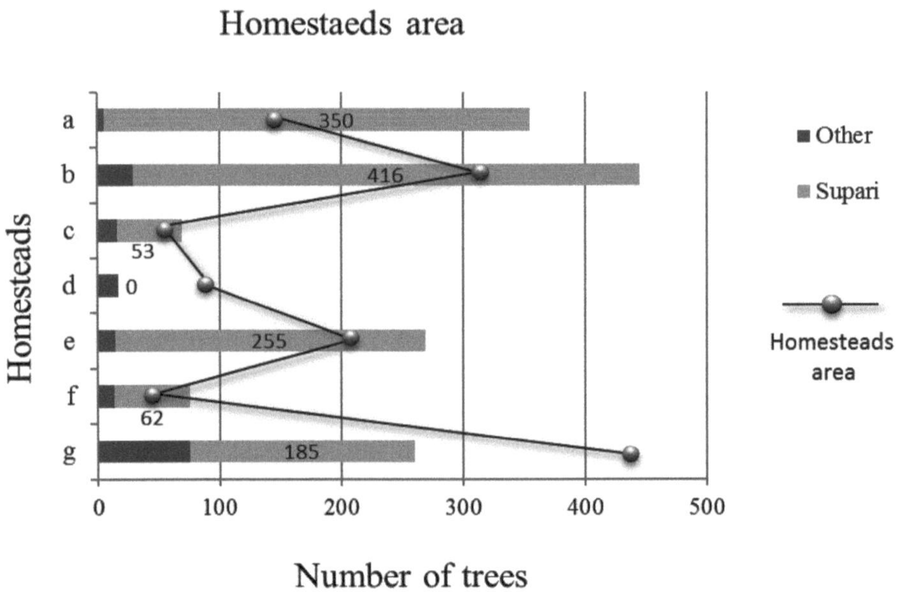

Fig. 13.7 Betel nut trees and other trees in relation to homestead size

This study identified 193 tree species, 44 of which accounted for more than 50% of the total tree number. The DBH of four tree species, namely, mango, jackfruit, coconut, and garjan, was greater than 30 cm. These species provide fruit and other livelihood necessities, but garjan trees were kept for conservation purposes. Although most of the homesteads were dominated by betel nut trees, 12 other tree species were also recorded. The proportion of betel nut trees seemed to be related to homestead area (as larger homesteads had more trees), side business as an occupation of the householders, and the level of education of the householders.

Most of the homesteads' plots are covered by a betel nut canopy, with shrub and herb layers in the understory. However, many other species, including vegetables, were found in the sunny front yards. There is the potential for local farmers to plant and grow various types of plant species in their homesteads, as homesteads d and g, which were settled after the 1990s, showed multiple production activities.

The findings reveal that betel nut planting is growing tremendously, as it ensures high income. This trend would decrease the biodiversity of the shrub and herb layers, as there is limited space under the betel nut canopy. It is important to develop an appropriately designed homestead production system that includes many trees, crops, and other production components. This would ensure food production, food and nutrient security, and high incomes, as well as provide a good living environment.

References

CWBMP (2011) Coastal and wetland biodiversity management at Cox's Bazar and Hakaluki Haor (CWBMP) [Online]. Available: http://www.doe-bd.org/cwbmp/. Accessed on May 2011

Iftekhar MS (2006) Forestry in Bangladesh. Over J For 104(3):148–115

Kabir ME, Webb EL. 2008. Can homegardens conserve biodiversity in Bangladesh? Biotropica 40 (1): 95–103

Mannan MA (2006) Plant biodiversity management at Teknaf Peninsula ECA. Coastal & Wetland Biodiversity Management Project, Cox's Bazar Site Office. 41:7–8

Muhammed N, Koike M, Haque F, Miah MD (2008) Quantitative assessment of people oriented forestry in Bangladesh: a case study from Tangail Forest Division. J Environ Manag 88(1):83–92

Rahman MM, Furukawa Y, Kawata I, Rahman MM, Alam M (2006) Role of homestead forests in household economy and factors affecting forest production: a case study in southwest Bangladesh. J For Res 11(2):89–97

Uddin MZ, Alam MF, Hassan MA (2012) Diversity in angiosperm flora of Teknaf Wildlife Sanctuary, Bangladesh. Bangladesh J Plant Taxon 20(2):145–162

Chapter 14
The Influences of Ethnicity and the Distance of Settlements from Forests on Homestead Production Systems and Ecosystem Services

Md. Abiar Rahman

Abstract Forest degradation is a major problem on the Teknaf peninsula where forest resources have shown a significant decline in recent years. A homestead in Bangladesh is made up of various components, including timber, fruit, vegetables, poultry, and livestock species. A homestead ecosystem has the potential to provide various services and products. Four distinct ethnic communities live in and around forest areas on the Teknaf peninsula, and homestead production systems and ecosystem services may vary across these communities. Moreover, ecosystem services may differ between forest and non-forest villages. This chapter presents the findings of an investigation of homestead production systems and ecosystem services within different ethnic communities. Variations in ecosystem services across forest and non-forest villages were also compared. The homesteads of households belonging to the Rakhine ethnic group were found to be relatively older, whereas those within the Chakma community were larger. Incomes generated from homesteads were higher for Chakma households, followed by those of households within the Bengali and Rakhine communities. Plant diversity was higher within Rakhine homesteads. The two forest villages in the study had large numbers of fruit species, whereas the density of timber species was high in the two non-forest villages. Native plant species predominated in the forest villages, whereas exotic species had been planted in homesteads in the non-forest villages. A conclusion of the study was that, because homesteads on the Teknaf peninsula supply various services and products such as fruit, timber, vegetables, meat, and eggs, designing a new homestead production system could maximize the ecosystem services provided by a homestead.

Keywords Homestead • Ethnicity • Forests • Ecosystem services • Population settlement • Plant diversity

---

M.A. Rahman (✉)
Department of Agroforestry and Environment, Faculty of Agriculture, Bangabandhu Sheikh Mujibur Rahman Agricultural University, Gazipur, Bangladesh
e-mail: abiar@bsmrau.edu.bd

M. Tani, M.A. Rahman (eds.), *Deforestation in the Teknaf Peninsula of Bangladesh*, DOI 10.1007/978-981-10-5475-4_14

14.1 Introduction

A homestead is arguably the most important production unit in Bangladesh. Around 20 million households maintain home gardens in the country, which has an urban population of about 25.36 million and a rural population of 21.90 million (Kabir and Webb 2008). These gardens cover an area of 0.27 million ha, which is equivalent to 2% of the country's total land mass, and provide approximately 75% of the nation's forest products (Kabir et al. 2016). It is noteworthy that whereas 25% of forest cover in a country is required to maintain an ecological balance, only 6–8% of the total land area of Bangladesh is forested. Massive degradation and depletion of natural forest resources have caused a number of meteorological and health hazards, posing difficult challenges for maintaining ecological sustainability (Foysal et al. 2013). Homestead-based agroforestry can play a promising role in counterbalancing the catastrophic events of climate change as well as maintaining a sound environment. The conservation of cultivated plants within Bangladesh's homestead gardens not only preserves a vital resource for humankind, but as a sustainable source of fruits and vegetables, these gardens also play an important role in safeguarding households' food security. However, the management of the traditional homestead garden has evolved as a response to many factors: tradition, culture, ethnicity, previous experience, education, economic and environmental factors, as well as personal preferences (Muhammed et al. 2013). Understanding the key driving forces that influence farmers' decisions relating to investments in homesteads is important not just for exploring human-environment linkages but also for potentially improving livelihoods through more effective management strategies.

Ecosystem services are benefits that people receive from nature and include, for example, food production, the provision of clean water, and climate regulation, as well as opportunities for cultural, spiritual, and recreational experiences (CCI and BirdLife International 2011). In recent years, a lot of flora and fauna have either become extinct or have reached an endangered state because of human activities and environmental changes. As a result, ecosystem services have also changed markedly, and many are in a reduced or degraded state. Measuring and monitoring ecosystem services can have many positive outcomes. It can promote better planning decisions that support both biodiversity conservation and ecosystem service delivery; help to identify and inform management strategies to enhance economic sustainability and human well-being; provide information on additional benefits accruing from traditional approaches to biodiversity conservation; identify those affected by land use management decisions, thereby helping to spread costs and benefits more fairly among stakeholders; and provide information to raise awareness and build support within the public and the government for evidence-based policy and management decisions (CCI and BirdLife International 2011).

Bangladesh is a country that aspires to provide fundamental rights for all communities across ethnic and religious differences. There are 45 ethnic communities living in different locations in Bangladesh. It is believed that homestead production systems, including the selection of plant species and their arrangements, largely depend on the preferences of these ethnic groups and that they differ from community to community. On the Teknaf peninsula, which is in the southeastern corner of Bangladesh, there are four distinct ethnic groups, namely, Bengali, Rakhine, Chakma, and Rohingya, who are living either in clusters or sporadically (Tani et al. 2011). Homesteads in this region evidence high diversity and large numbers of plant species (Asahiro et al. 2013; Nath et al. 2015).

As described in previous chapters, human settlements have been established in and around the reserved forest area as well as far from forests. As a result of overexploitation, forest resources and diversity have undergone a dramatic decline during the last couple of decades on the Teknaf peninsula. Various homestead components such as trees, crops, livestock, and poultry may help to reduce pressure on forests by providing various products and services. We can assume that homestead production systems and services vary according to the distance of homesteads from forests and according to different practices among ethnically distinct communities on the Teknaf peninsula.

14.2 Methodology

Two separate studies were designed to investigate homestead production systems and ecosystem services on the Teknaf peninsula in 2015.

For the first study, a total of 100 households belonging to four ethnic communities were randomly selected from different settlements on the peninsula (Fig. 14.1). Twenty-five households from each community were surveyed between March 2015 and May 2015 using a structured interview schedule. Along with focus group discussions held with members of the households, physical monitoring was also conducted and measurements of various plants were taken.

For the second study, 50 households from two forest villages (Rangikhali and Rajarchara) and 50 households from two non-forest villages (Dail Para and Hariakhali) on the Teknaf peninsula were randomly surveyed (Fig. 14.2). The survey was conducted from September to November 2015 using a structured interview schedule. For this study too, focus group discussions were held with members of households, and physical monitoring was conducted and measurements of various plants were taken.

Similar methods to those described in Chap. 12 were used to measure and assess homestead production systems, tree performances, and household incomes. Species richness was measured using the formula $D = s/\sqrt{N}$ where D denotes species richness, s is the number of different species represented in a homestead, and N denotes the total number of individual organisms in a homestead. Species diversity was measured using the Shannon index, $H = -\Sigma Pi\text{Ln}Pi$, where Pi denotes the number of individuals of one species/total number of individuals in the samples. The relative prevalence of species (RP) was calculated as the number of species in a homestead/total number of species in a homestead × 100.

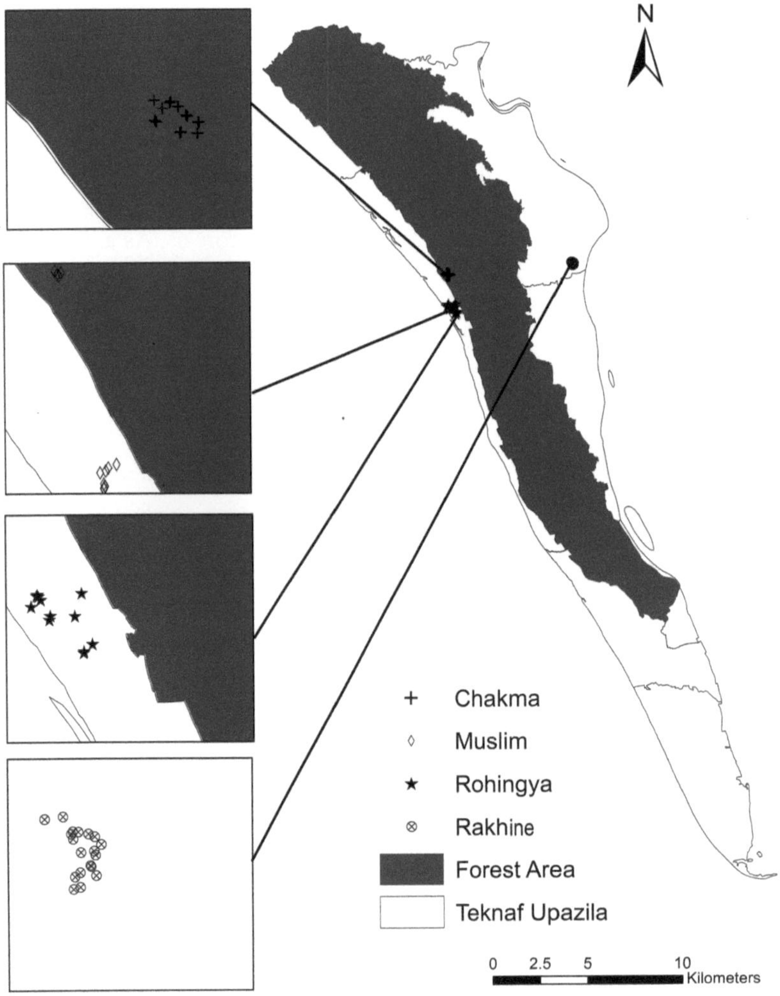

Fig. 14.1 Survey locations of different ethnic communities in the Teknaf peninsula

14.3 Homesteads of Households from Different Ethnic Communities

Table 14.1 shows the ages and areas of homesteads within the four ethnic communities under investigation. It is evident from the table that Rakhine respondents had the oldest homesteads (about 55 years), followed by Chakma respondents (36.5 years), which was closely followed by those of Bengali respondents (27.5 years).

By contrast, the homesteads of Rohingya respondents were recent (established 6 years ago). Homestead areas were greatest among Chakma households (0.24 ha),

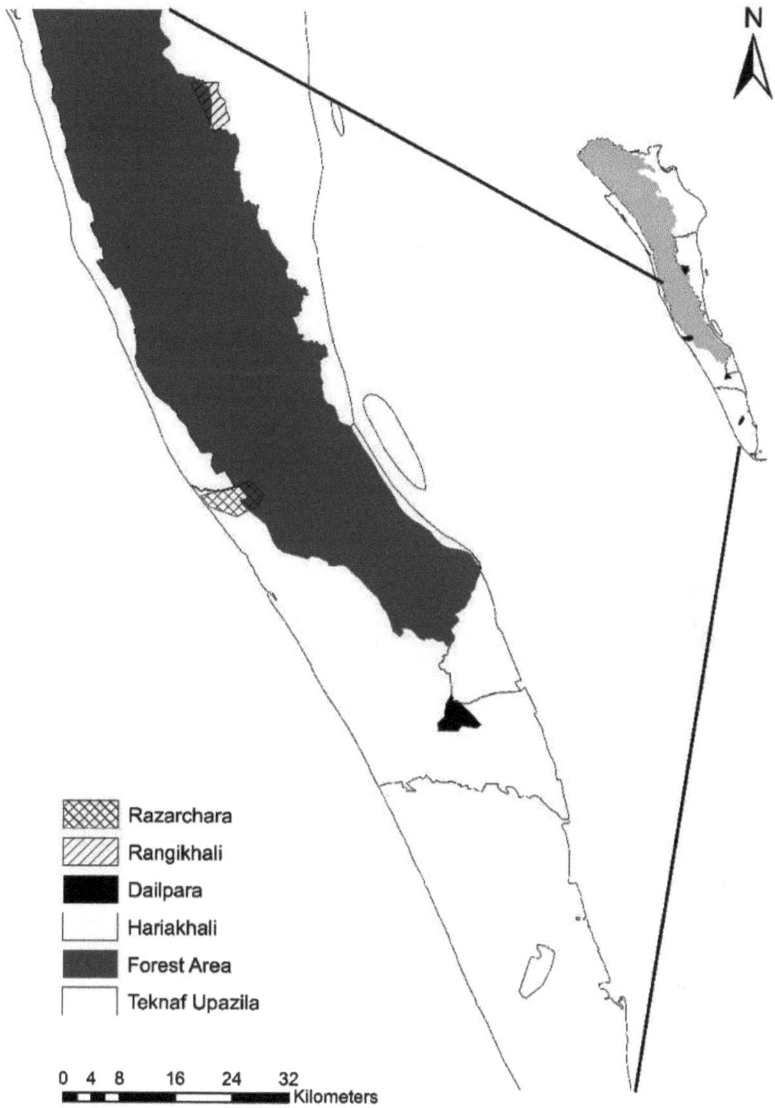

Fig. 14.2 Study locations of forest and non-forest villages in the Teknaf peninsula

followed by those of Bengali households (0.19 ha), which were statistically similar to the homestead areas of Rakhine households (0.16 ha). Homestead areas were smallest (0.07 ha) among Rohingya households.

Table 14.2 depicts total annual incomes of households belonging to different ethnic groups within each community, as well as homestead incomes and their contribution to the total annual incomes of these households. The total annual incomes of Rakhine households were found to be the highest (268.7 thousand BDT), followed

Table 14.1 Age and area of homesteads of different communities

Ethnicity	Homestead age (year)	Area (ha)
Rakhine	55.1 (±3.01)a	0.16 (±0.02)b
Bengali	27.5 (±5.4)b	0.19 (±0.03)b
Rohingya	6.0 (±1.48)c	0.07 (±0.02)c
Chakma	36.5 (±6.3)b	0.24 (±0.07)a
Mean (n = 100)	31.3 (±2.98)	0.17 (±0.04)
F test (P)	<0.001	<0.05

Table 14.2 Total annual income and contribution of homestead income to the annual income of households of different ethnic groups

Ethnicity	Total annual income (1000 BDT)	Homestead annual income (1000 BDT)	Contribution (%)
Rakhine	268.7 (±35.1)a	16.1 (±3.7)ab	6.0
Bengali	149.7 (±14.9)b	17.9 (±3.4)bc	12.0
Rohingya	109.4 (±20.6)b	7.0 (±3.1)c	6.4
Chakma	123.1 (±16.6)b	21.6 (±5.5)a	17.6
Mean	177.7 (±16.8)	15.7 (±4.6)	10.5
F test (P)	<0.05	<0.05	

Notes: The figures in parentheses indicate standard deviation (SD) values. Data for 25 households within each community were analyzed. Means followed by different letters differed significantly at the 5% level according to the results of Duncan's multiple range test (DMRT)

by those of Bengali households, which were statistically similar with the remaining ethnic groups. Chakma respondents recorded the highest homestead annual incomes (21,600 BDT), whereas those of Rohingya respondents were the lowest (7000 BDT). If we look at the percentage contribution of homestead incomes to annual household incomes among the different ethnic groups, it is evident that this contribution was highest among Chakma households (17.6%), followed by Bengali households (12%). By contrast, Rohingya and Rakhine respondents reported the lowest contributions of homestead incomes to their annual household incomes.

14.4 A Comparison of Homesteads in Forest and Non-forest Villages

A large number of fruit trees occurred within homesteads located in both forest and non-forest villages, followed, in terms of numbers, by timber and medicinal plant species. The number of plant species was higher in forest villages than in non-forest villages, possibly because of the larger areas of homesteads in the former. A higher density of fruit trees was found in forest villages, while a higher density of timber species was found in the non-forest villages (Table 14.3). Households in forest villages had easy access to timber and other forest products and may not therefore have been particularly interested in growing timber species. On the other hand, households in non-forest villages had fewer options for procuring timber and fuelwood. They therefore planted large numbers of timber trees in their homesteads.

Turning to livestock and poultry, hens were prevalent in both forest villages (80%) and non-forest villages (98%). In forest villages, goats, cows, ducks, and buffaloes were also reared in some homesteads, whereas cows, goats, and pigeons were kept in homesteads located in non-forest villages (Table 14.4).

While deforestation is a major problem on the Teknaf peninsula, the ecology of this area remains suitable for cultivating various plant species. As in other parts of Bangladesh, exotic native plant species are found on the Teknaf peninsula. More than 60% of homesteads in the non-forest villages in this study contained exotic species, whereas the percentage of homesteads with these species in the forest villages was only 34% (Table 14.5). Native plant species dominated homesteads in

Table 14.3 Plant resources found in forest and non-forest villages in the Teknaf peninsula

	Forest village	Non-forest village	Forest village	Non-forest village
Species	Number/homestead		Density (Number/ha)	
Timber	20	16	185	197
Fruit	61	31	563	394
Medicinal and others	3	1	28	13

Table 14.4 Livestock and poultry resources found in forest and non-forest villages in the Teknaf peninsula

	Forest village	Non-forest village
Item	Homestead (%)	
Cow	20	14
Buffalo	2	0
Goat	26	6
Hen	80	98
Duck	4	0
Pigeon	0	2

Table 14.5 Incidence pattern of plant species found in homesteads of forest and non-forest villages in the Teknaf peninsula

	Forest village	Non-forest village
Species	Homestead (%)	
Exotic	34	64
Cotton tree	36	4
Bamboo	50	4
Betel nut	64	56
Bilimbi	36	4
Painna gola	14	6
Hibiscus	18	6

forest villages as opposed to those in non-forest villages. Some of the most common native plant species found on the Teknaf peninsula include *tula*, bamboo, betel nut, *bilimbi*, *painna gola*, and hibiscus.

14.5 Ecosystem Services from Homesteads

Biodiversity is an important asset that aids human survival and well-being in many ways. The benefits derived by humans from nature are referred to as "ecosystem services." They can be divided into processes (e.g., soil formation) underpinning services (e.g., crop production), which in turn provide goods (e.g., food), often in conjunction with other inputs (e.g., labor). Ecosystem services can be valued in monetary (market and nonmarket) as well as nonmonetary terms to indicate their contribution to economic, health, and social well-being (CCI and BirdLife International 2011). Ecosystem services have different impacts on practitioners, depending on who they are, where they live, and when they use the services. However, these services and their impacts are often overlooked.

The different ethnic groups of Bangladesh, with their culturally vibrant lifestyles, have significantly enriched the overall cultural identity of the nation. For centuries, Bangladesh has been home to different ethnic groups. In fact, 45 smaller groups of indigenous people covering about 2% of the total population have been living in different pockets of the country's hilly zones and in some plain areas. There are four main ethnic groups on the Teknaf peninsula, namely, the Bengali, Rohingya, Rakhine, and Chakma groups. Their historical backgrounds, economic activities, social structures, and religious beliefs and festivals make them distinctive in addition to having a key influence on homestead resource management.

Table 14.6 illustrates the number of tree species within individual communities as well as the mean number of trees per homestead, tree diversity, and homestead areas across the different ethnic groups. As indicated by the table, Chakma homesteads had the greatest tree diversity (19 species), whereas tree diversity in

Table 14.6 Tree species and tree diversity in homesteads of different ethnic groups

Ethnicity	Number of tree species	Number of trees per homestead	Shannon–Wiener diversity index	Evenness (E)
Rakhine	17 (±5.5)	123 (±22.2)a	2.01 (±0.14)a	0.61 (±0.04)a
Bengali	16 (±3.5)	142 (±23.8)a	1.21 (±0.12)b	0.54 (±0.04)b
Rohingya	14 (±1.8)	42 (±6.6)b	1.32 (±0.13)b	0.41 (±0.06)b
Chakma	19 (±5.7)	131 (±33.0)a	1.81 (±0.14)a	0.59 (±0.07)a
Mean (n = 100)	17 (±4.7)	110 (±18.7)	1.59 (±0.16)	0.54 (±0.06)
F test (P)	n.s.	<0.05	<0.05	<0.05

Notes: Figures in parentheses indicate standard deviation (SD) values. Twenty-five households within each community were analyzed. Means, followed by different letters, were significantly different at the 5% level according to the results of Duncan's multiple range test (DMRT)

Rohingya homesteads was lower (14 species). However, in terms of the greatest number of trees, households of Bengali ethnicity ranked highest with an average of 142 trees per homestead. This figure is statistically similar to that of the Chakma and Rakhine homesteads which had, on average, 131 and 123 trees, respectively. By contrast, Rohingya households had the lowest average number of trees (42) per homestead. The highest values for the diversity index, as well as evenness within homesteads, were obtained for households within the Rakhine community and were statistically similar to those of households belonging to the Chakma ethnic group, followed by those in the Rohingya, a Bengali group, respectively.

Table 14.7 shows the relative prevalence, species richness, and diversity index of plant species grown in forest and non-forest villages. Whereas non-forest villages evidenced richness and diversity relating to timber species, the richness and diversity of fruit and medicinal plant species were more evident in forest villages (Table 14.7). Households in non-forest villages are evidently inclined to plant fast-growing species to procure fuelwood and timber supplies. Therefore, they are interested in planting exotic species. On the other hand, people in forest villages plant more fruit trees to obtain other benefits, as they are easily able to access fuelwood and timber supplies from nearby forests.

A homestead provides a wide range of services and supplies. Fruit, timber, vegetables, meat, milk, eggs, and fuelwood are some of the common products that households obtain from homesteads, although these supplies are insufficient to meet households' demands (Table 14.8). No fodder or fish were found in homesteads either in forest or non-forest villages on the Teknaf peninsula.

Table 14.7 Plant species diversity found in homesteads of forest and non-forest villages in the Teknaf peninsula

	Relative prevalence		Species richness		Shannon index	
Timber species	Forest village	Non-forest village	Forest village	Non-forest village	Forest village	Non-forest village
Timber	21.57	35.82	13.77	18.14	0.34	0.37
Fruit	73.06	64.77	47.21	31.75	0.23	0.28
Medicinal	2.78	1.99	1.80	0.98	0.10	0.08

Table 14.8 Comparison of forest and non-forest villages in terms of services obtained from homesteads of the Teknaf peninsula

	Forest village	Non-forest village
Services	Supply (%)	
Fruit	86	86
Timber	24	32
Vegetable	48	42
Meat	76	96
Milk	6	2
Egg	74	94
Fuelwood	14	44

In general, the cultivation of home gardens in Bangladesh has entailed the ongoing practice of inherited traditions, which have evolved through centuries of experience, observation, and trial and error (Islam et al. 2015). Plant species and diversity in homesteads vary from place to place and are largely influenced by ecological and socioeconomic factors. They also vary among homesteads, even those with similar ecologies, maintained by households within the same socioeconomic group, depending upon individual needs and preferences. Older homesteads may contribute to increased species richness within a home garden. It has been widely posited that for subsistence-oriented home gardens, there is a positive correlation between homestead size and home garden species richness (Igwe et al. 2014; Sabastian et al. 2014). Many other studies have found that there is no relationship between homestead size and species diversity (Saikia et al. 2012; Aworinde and Erinoso 2013). However, the positive relationship found by one study between homestead size and species richness in the majority of home gardens in southwestern Bangladesh (Kabir and Webb 2008) supports the generalized hypothesis relating to subsistence-oriented homesteads.

After agriculture, home gardens are ranked as the second most important source for meeting subsistence needs for many Bangladeshi households. A vast majority of rural households in Bangladesh that cultivate crops on their land remain unemployed for a considerable portion of the year because of the seasonality of production activities and labor requirements. Homestead farming is the best solution to this unemployment situation through the cultivation of both vegetables and fast-growing fruits, enabling household members to remain employed throughout the year (Miah and Hussain 2001). Studies have found that over the last few decades, small-scale homestead activities have become the most significant income-generating activities and ecosystem services for poor households. Species richness in home gardens has a significant positive correlation with households' economic conditions. The economic statuses and patterns of households are often reflected in the plant diversity of their home gardens (Aworinde and Erinoso 2013; Riu-Bosoms et al. 2014), although the exact nature of this relationship is not known. The relationship between species composition and household incomes has been hypothesized to be positive, as richer households can be expected to grow more crop varieties than poorer households (Linger 2014). The augmentation of income increases the likelihood that farmers will adopt high-quality planting materials relating to diverse species. Farmers who have more land (Igwe et al. 2014), have higher incomes (Sabastian et al. 2014), and spare family labor (Abebe 2005) can afford to plant and maintain more trees for home garden production (Linger 2014). Our results revealed that the richness of plant species in home gardens would increase with increased household landholdings and incomes.

14.6 Conclusions

It is important to quantify ecosystem service(s) provided by a site in its current state, as this may facilitate household decision-making. On the Teknaf peninsula, ecosystem services provided by homesteads can help to reduce pressure on forest resources within a highly degraded forest ecosystem. Homestead ecosystem services can support the conservation or restoration of a site. Data from this study revealed that homesteads within the Rakhine community contained many ornamental plants, which provide a good living environment as well as esthetic value. On the other hand, Chakma homesteads included a large number of vegetables that ensure higher household incomes. The study also found a high level of plant species diversity on the peninsula. There was a significant degree of variation between homesteads in forest and non-forest villages in terms of their homestead production systems, species diversity, and services. Whereas homesteads in forest villages evidenced dense and diversified populations of fruit trees, timber species predominated in homesteads in non-forest villages.

Homesteads clearly offer an option that entails the rational use of small areas of land, along with the provision of various services. These services include livelihood support as well as soil and water conservation, biodiversity conservation, energy supplies in rural areas, afforestation, carbon sequestration, landscaping, and environmental improvement. Thus, the development of an economically viable and ecologically sustainable homestead production system should be one of the policy objectives relating to the conservation of natural resources and their sustainable management in Bangladesh and in other tropical and subtropical regions.

References

Abebe T (2005) Diversity in homegarden agroforestry systems of southern Ethiopia. Ph.D. dissertation, Wageningen University, Wageningen, p 153

Asahiro K, Takada M, Rahman MA, Tani M (2013) Tree census and household lifestyle in homestead of Bangladesh – a case study in the west coast of Teknaf. In: Proceedings of the International conference on environmental aspects of Bangladesh. University of Dhaka, Dhaka

Aworinde DO, Erinoso SM (2013) Relationship between species composition and homegarden size in Odeda Lga of Ogun State Nigeria. Bayero J Pure Appl Sci 6(2):10–18

CCI and BirdLife International (2011) Measuring and monitoring ecosystem services at the site scale. Cambridge Conservation Initiative and BirdLife International, Cambridge

Foysal MA, Hossain ML, Rubaiyat A, Hasan MB (2013) Economics of homestead forestry and their management activities at Fatickchari Upazila of Chittagong District, Bangladesh. Agric For Fish 2(4):161–176. doi:10.11648/j.aff.20130204.12

Igwe K, Agu-Aguiyi F, Nwazuruoke G (2014) Social and economic implications of home gardening on the livelihood of farm households in Abia State, Nigeria. Dev Country Stud 4(1):66–71

Islam M, Dey A, Rahman M (2015) Effect of tree diversity on soil organic carbon content in the homegarden agroforestry system of North-Eastern Bangladesh. Small Scale For 14:91–101

Kabir ME, Webb EL (2008) Can homegardens conserve biodiversity in Bangladesh? Biotropica 40(1):95–103

Kabir ME, Rahman MM, Rahman MAU, Ando K (2016) Effects of household characteristics on homegarden characteristics in Kalaroa Upazila, Satkhira District, Bangladesh. Small Scale For:1–19

Linger E (2014) Agro-ecosystem and socio-economic role of homegarden agroforestry in Jabithenan District, North-Western Ethiopia: implication for climate change adaptation. Springer Plus 3:1–9

Miah MG, Hossain MK (2001) Study of the indigenous knowledge on the homestead forestry of Narsingdi region, Bangladesh. S Asian Anthropogist 1(2):129–135
Muhammed N, Masum MFH, Hossain MM (2013) Economic dependence of rural people on homestead forestry in Mymensingh, Bangladesh. J For Res 24(3):591–597
Nath TK, Aziz N, Inoue M (2015) Contribution of homestead forests to rural economy and climate change mitigation: a study from the ecologically critical area of Cox's bazar—Teknaf Peninsula, Bangladesh. Small Scale For 14(1):1–18
Riu-Bosoms C, Calvet-Mir L, Reyes-Garcıa V (2014) Factors enhancing landrace in situ conservation in home gardens and fields in Vall de Go'sol, Catalan Pyrenees, Iberian Peninsula. J Ethnobiol 34(2):175–194
Sabastian G, Kanowski P, Race D, Williams E, Roshetko JM (2014) Household and farm attributes affecting adoption of smallholder timber management practices by tree growers in Gunungkidul region, Indonesia. Agrofor Syst 88(2):257–268
Saikia P, Choudhury BI, Khan ML (2012) Floristic composition and plant utilization pattern in homegardens of Upper Assam, India. Trop Ecol 53(1):105–118
Tani M, Rahman MZ, Rahman MA, Asahiro K, Akhter S (2011) Deforestation by daily activities in the Teknaf Peninsula, Bangladesh. Proceedings of the International Conference on Environmental Aspects in Bangladesh (ICEAB) 2011: 73–75

Part IV
Conclusions

Chapter 15
Conclusions

Masakazu Tani

Abstract This book is a culmination of 5 years of investigation of deforestation on the Teknaf peninsula, Bangladesh. This research was launched based on the premise that severe deforestation on the Teknaf peninsula was a socially constituted phenomenon that was particularly affected by poverty in this region. The studies presented here have investigated two major themes: (1) the causes of deforestation and (2) possible reforestation strategies.

Keywords Causes of deforestation • Reforestation strategies • Poverty connection • Recomendations

Regarding the first theme, this study found three major causes of deforestation in the study area, namely, betel leaf cultivation, fuelwood collection, and human encroachment into the forest, with each cause being related to poverty to varying degrees. Betel leaf is the major cash crop of the Teknaf peninsula and is widely cultivated because cheap shading materials and good edaphic and environmental conditions are readily available. A series of studies was conducted to determine how betel leaf cultivation is affecting forest growth on the peninsula. Betel plants are highly susceptible to pests and diseases, and an outbreak of a disease may ruin all of the plants within a patch. Therefore, a betel farmer needs to exercise great care, repeatedly applying agricultural chemicals to control and monitor conditions inside a *pan boroj*. Betel leaf cultivation is also labor intensive. Besides such applications of agricultural chemicals, these plants require frequent watering, at least twice a day. Because of the lack of automatic sprinklers, most betel farmers manually transport water from a source to the plot each time they water the plants. Although betel leaf cultivation is risky and entails high costs, farmers are nevertheless motivated to grow it, because its income potential per unit area is higher than that of rice. During the course of our study, we observed two activities relating to betel leaf cultivation

M. Tani (✉)
Department of Environmental Design, Faculty of Design, Kyushu University, Fukuoka, Japan
e-mail: tani@design.kyushu-u.ac.jp

M. Tani, M.A. Rahman (eds.), *Deforestation in the Teknaf Peninsula of Bangladesh*, DOI 10.1007/978-981-10-5475-4_15

that affect forests: (1) felling trees to construct shade-providing *pan boroj* structures and (2) clearing forests to establish cultivation plots. Betel leaf fields occupy approximately 5% of the forest area, and *pan boroj* construction consumes approximately 4% of the forest materials produced annually. Therefore, betel leaf cultivation appears to be a major cause of forest depletion. Although betel leaf cultivation can be more profitable per unit area than rice cultivation, betel leaf farmers are relatively poor (although not the poorest members of society) and less educated, on average, than other groups, and in some cases they are illegal forest dwellers. They mainly belong to the lower socioeconomic tier of farmers who attempt to compensate for their lack of resources, including the lack of arable land, by engaging in less stable but potentially more productive farming practices such as betel leaf cultivation. This is partly possible because the materials needed to engage in betel leaf farming can be freely obtained from the forest.

Fuelwood is the single greatest source of household energy in the Teknaf peninsula. The greatest proportion of fuelwood comes from forests. Three groups of people that are involved in fuelwood collection have been characterized, namely, purchasers, collectors, and sellers. The last two groups are directly involved in felling trees and shrubs. People who physically fell and collect trees in the forest are mostly poor, and they are generally poorer than betel leaf farmers. Fuelwood collectors typically possess few resources. They may have no stable occupation and little land other than their homestead. Felling trees for sale may be their only means of survival, although such activities are prohibited by law. This indiscriminate tree felling in the restricted area critically affects forest regeneration on the Teknaf peninsula. However, a strict ban on felling trees may be difficult to enforce. A lack of personnel hinders the Forest Department's efforts to comprehensively control the reserve forest. More importantly, however, such a ban would have an immediate negative effect on the livelihoods of many residents who would lose their only means of subsistence.

People living in the restricted area of the Teknaf Wildlife Sanctuary (TWS) include forest villagers and original settlers. Along with these legal residents, a large number of other settlers were found to be encroachers. The local administration is unable to control them given the sociopolitical and humanitarian issues involved. It is likely that these encroachers would further degrade the condition of already damaged forests. These encroachers are generally poor and landless. They engage in tree felling and small-scale farming, including betel leaf cultivation, by clearing the forest for farming plots. The influx of encroachers into the forest is continuing, and it has accelerated in recent years, partly because of an increase in the number of impoverished people living around the forest.

The presence of a large number of Rohingya refugees on the Teknaf peninsula is a critical issue that might not be resolved in the immediate future. The Rohingya are a Muslim minority group in Myanmar, whose government has denied their nationality, because it regards the Rohingya as illegal immigrants from Bangladesh. Because of their oppressive treatment in Myanmar, Rohingya refugees have been fleeing in large numbers to Bangladesh during the last few decades. Consequently, a large number of Rohingya refugees have settled in and around the TWS and are mainly engaged in farming and fishing. Along with the local poor, they also collect fuelwood. Although it is widely believed that the Rohingya are primarily responsible for deforestation on the Teknaf peninsula, the findings of this study show that they are not the major factor influencing deforestation. Rather, this issue should be viewed as one related to poverty.

The second theme of this research investigated methods by which forests on the Teknaf peninsula could be regenerated. Social forestry (SF) programs and homesteads were assessed in terms of their reforestation potential.

The Forest Department started an SF program in 1981/1982 with foreign support. SF has been reported to be successful in conserving forests, particularly in the northern part of Bangladesh, with the active participation of local people. In Teknaf, SF was started in 2003. SF conditions in the Teknaf and Shilkhali ranges were examined, and 37 active SF plots were identified. *Gmelina arborea* and *Acacia* spp. are the most dominant tree species in the studied SF plots. It was found that some plots are not managed properly, whereas 11 plots are in good condition. An actor analysis showed that the Forest Department is the most powerful and influential actor in SF. They gained power via the bureaucratic forest management and policies of Bangladesh. For the success of SF, it is important to empower the local people by providing proper logistical support.

Homesteads are widely found in Bangladesh like many other tropical countries, where fruit and timber species, vegetables, livestock, and poultry are commonly found. Even landless households have homesteads, which are the centers of livelihood activities. Homesteads are larger in Teknaf than in other parts of Bangladesh. Horizontal and vertical arrangements of the vegetation show no particular pattern of species arrangement. However, the species are diversified. Betel nut is the most dominant tree species, and it makes a large contribution to annual household income. In general, the proportion of betel nut trees is related to homestead area. Although people plant a lot of betel nut trees in their homestead, they do not manage the trees very well. Vegetables, fruits, poultry, and livestock are not attended closely indicating poor management of homestead production system. Moreover, a sole interest in betel nut trees would decrease biodiversity, especially for shrub and herb species. It is important to develop a good homestead production system with an appropriate design that incorporates many trees, crops, and other production components. Such a practice ensures food, nutrients, energy, ecological security, and a good living environment, and an increased provision of fuelwood from homesteads may relieve some pressure from the forest.

15.1 Poverty Connection

Betel leaf cultivation, fuelwood collection, and encroachment into the forest negatively affect the state of forest in Teknaf. However, it is not simple to prevent further deforestation from advancing by banning these activities, because they are an integral part of people's, particularly the poorer segments of the population, livelihoods.

People engage in such activities to survive in a less favorable physico-social environment.

Betel leaf cultivation is a means of effectively using severely limited land resources for productive purposes, and, therefore, it is a highly adaptive practice. There is not a viable alternative to individual farmers throughout the region as a whole. Without it, the livelihoods of a large portion of the local population would immediately collapse.

Fuelwood collection, which often occurs in the restricted forest of the TWS, is often the only means by which the poorest segment of the population can make a living. These people have virtually no properties and resources upon which to rely; thus, freely available fuelwood is of great value.

15.2 Recommendations for Future Studies

Our study has led to a greater understanding of deforestation in Teknaf. Future research will be required to further mitigate this environmental/social disaster and to improve the living environment of the Teknaf peninsula.

These future studies should focus on regional planning and resource management. The current deforestation problem, as seen in this volume, seems to be caused by various types of indiscriminate uses. Many types of human activities negatively affect the forest in this area, including the collection of materials for shading betel leaf plots, opening such plots in the forest area, fuelwood collection, and encroachment into the forest area. These activities are mostly illegal, because they all use supposedly protected resources in the reserve forest. However, people engaged in these activities are generally poor, and access to forest resources is critical for their survival. Therefore, a total ban on the use of resources of the reserve forest may not be a viable solution to the problem.

Controlling the forest resource use activities in this area would not be practical either. As deforestation in Teknaf has already become a serious environmental problem, the forest in this area will be irreversibly lost in the near future if countermeasures are not implemented soon. Such an outcome would be an environmental and social disaster, because a large portion of the current population would lose their only means of survival.

Therefore, this area urgently requires further research to identify ways to sustain nature and people in this area by using and regenerating resources indefinitely. The forest is indispensable for sustaining human livelihoods in this area. Thus, the forest should not be sacrificed for the people, but at the same time, the people, particularly the poor segments of the population, should not be sacrificed to preserve the forest.

A first step in such research would be to assess the balance between the environment and people's needs. The productive capacity of the local forest and local demands for trees should be quantitatively characterized. Sustainable ways of planning based on a gross balance between the demand and supply should also be studied.

A basic scheme for this planning may be a zoning approach. Zones may include betel leaf cultivation plots, areas that provide materials for *pan boroj* construction, fuelwood collection areas, controlled settlement areas, and forested areas. Zoning for betel leaf cultivation by restricting the use of forest resources as shading materials may be a viable option. New production technologies that use alternative shading materials should be investigated. Moreover, the establishment of new betel leaf fields, particularly in the TSW, must be restricted. An in-depth study on betel leaf farmers should be conducted to analyze the socioeconomics, profitability, productivity, and marketing systems of betel leaf cultivation, as well as their impacts on forests and other natural resources.

Teknaf forests are a prime source of fuelwood for the locality, the collection of which leads to deforestation. To protect the forest, fuelwood should be collected rotationally from different places to allow the forest to regenerate. Harvesting saplings should be restricted. People should be encouraged to use other energy sources, such as liquid petroleum gas, biogas, solar, and power. Homesteads can be a good option for supplying fuelwood.

Along with zoning, alternative ways of resource management are also important subjects of future research. The establishment of a buffer forest with the active participation of the local people would help to restrict people's access to the forest. The Forest Department started a long-rotation plantation using native and medicinal species. Local people should be more engaged in this activity. Moreover, the current drawbacks of the SF program on the peninsula should be analyzed to launch new and effective SF programs. Political and bureaucratic interference must be minimized by empowering the beneficiaries of SF. The experiences of other countries (e.g., India and Nepal) with SF should be considered. Sufficient and qualified manpower should be used to conserve forests. Strict systems for monitoring and controlling the use of forest resources by brickfields, sawmills, and timber merchants should be established. A co-management approach should be strengthened, and it should involve different sectors of society. Multidisciplinary involvement (forestry, agriculture, fisheries, livestock, environment, and socioeconomics) could ensure the sustainability of the program.

Homesteads can be another locus of an intensive study of resource production and conservation. Homesteads can be utilized more productively to contribute to the general household economy. Current household income from homesteads relies mainly on betel nut production. Unused land should be planted with multipurpose tree and shrub species to meet timber, fuelwood, fruit, and fodder demands. A good homestead production model should be developed to supply vegetables and fruits throughout the year. Cattle and poultry should be reared to supply protein and generate income. Women's participation in the homestead production system should be encouraged. The Forest Department should include homestead forestry in their forest conservation activities. Multistory agroforestry systems should be introduced to maximize resource production and forest conservation by selecting compatible tree and crop species.

Finally, poverty must be addressed in future research. Poverty is the biggest challenge to alleviating deforestation. Because it may be difficult to restrict poor people from entering the forest, alternative income-generating activities and employment opportunities should be explored by involving different stakeholders, including researchers, local leaders, policy makers, and administrators. Ecotourism would be a good way to increase the income of the local people. Small-scale businesses, such as plant nurseries, handicraft production, small shops, and other enterprises, can be established. Because deforestation in Teknaf is socially structured, multiple facets of this problem need to be addressed. Future studies may have a more comprehensive framework in terms of the scales and types of human activities. The scale should cover the ecological, economic, and social systems of the entire region. To this end, when determining the extent of human activity therein, all of the major stakeholders, primary livelihood activities, and institutions should be included.

Kolibri 360 GmbH
Ferdinand-Jühlke-Straße 7
99095 Erfurt, Deutschland
produktsicherheit@kolibri360.de